The Great Tree Story

The Great Tree Story

How Forests Have Shaped
Our World

Levison Wood

First published in Great Britain in 2025 by Gaia, an imprint of
Octopus Publishing Group Ltd
Carmelite House
50 Victoria Embankment
London EC4Y 0DZ
www.octopusbooks.co.uk

An Hachette UK Company
www.hachette.co.uk

The authorized representative in the EEA is Hachette Ireland,
8 Castlecourt Centre, Dublin 15, D15 XTP3, Ireland (email: info@hbgi.ie)

ISBN (Hardback) 978-1-85675-561-0
ISBN (Trade Paperback) 978-1-85675-562-7

A CIP catalogue record for this book is available from the British Library.

Typeset in 10.75/16.5pt Miller Text by Jouve (UK), Milton Keynes

Printed and bound in Great Britain.

3 5 7 9 10 8 6 4 2

This FSC® label means that materials used for the product have been responsibly sourced.

For all my Tree Teachers

Contents

CONTENTS

Introduction

———

One crisp autumn morning, when I was ten years old, I set off with my father on a walk in the local woods to gather conkers for a 200-year-old ritual. In those days, the childhood pastime of gleefully smashing horse chestnuts against each other marked the highlight of the school year in an age before computer games. As we rummaged through the carpet of fallen leaves in search of the finest and shiniest seeds to take home ready for the upcoming playground competitions, my dad suggested that we keep one to plant in a pot and watch it grow.

In my youthful rashness, I suspected this endeavour to be a slow and unexciting journey – not exactly a great spectator sport – yet I agreed, not fully grasping the significance of the proposal. Little could I have imagined how this small and simple act would lead to a remarkable transformation. Over the years, that little conker sprouted into a young sapling which quickly outgrew its pot and was eventually relocated to the front garden, where it spread its roots and started to reach for the skies.

As I grew older and ventured out into the world, I always enjoyed returning home to witness the tree's evolution – its branches strengthening, new leaves unfurling and the promise of conkers emerging with each passing season. I felt a deep-rooted connection as we grew together, even in our distance. Three decades later, it still stands there proudly, 20 feet tall, almost uprooting the wall next to which it was buried as a testament to our shared journey and the essence of home.

Trees are part of the family of life on Earth and it is about time they got the recognition they deserve. For me, walks in the woods fuelled an active imagination and a taste for adventure. A childhood spent outdoors inspired a thirst for stories of what lay beyond the fields and hedgerows of our green and pleasant land, and a curiosity about a time long ago, before the plough and tiller set to work. Back to an age when my forefathers lived among the trees, and from where they received their name.

Like many children, I came to understand what a true forest was through popular culture. I have vague recollections of reading *The Jungle Book*, watching black-and-white Tarzan films, and of course documentaries by Sir David Attenborough about the Amazon rainforest, which filled my head with visions of jaguars, parrots and monkeys. I was lucky to be blessed with a family that, despite living in a decaying city, had a healthy respect for nature. My mother read books to me like *The Magic Faraway Tree* by Enid Blyton, and my father, an avid nature man, took me to see the great trees of Sherwood Forest.

One clear night, I remember staring up at the sky and him telling me, 'There are more trees on Earth than there are stars in the Milky Way.' Three trillion, in fact. It occurred to me then that I should probably take a more active interest in them.

Both my grandfathers also encouraged a fascination with trees. On my father's side, Levison Wood (the Third) gave me a detailed description of his own adventures fighting the Japanese in the jungles of Burma during the Second World War. He told me about the perils of leeches and tigers, endless rains and gargantuan spiders, and that is before the jungle's grisly man-made dangers of pits filled with sharpened sticks, exploding booby traps and snipers tied to the canopy.

The jungle was no place for a boy, he told me, unless you had the good fortune, like Mowgli or Tarzan, to be raised by wolves or apes. Given his own teenage experiences among the trees, it is easy to forgive his reservations. Yet, despite his warnings, I kept dreaming of forests and what rewards these fantastical places might be concealing.

My other grandfather, Peter Curzon, claimed tenuous descent from Admiral John Benbow, whose great naval legacy included chopping down many of Britain's finest oak trees to turn into ships. Grandad Curzon took a different approach towards encouraging an appreciation for the outdoors. Every Saturday he would hide a 20-pence piece in his 'jungle' – a small thicket of rhododendron bush in his suburban garden. The jungle, he told me, was full of magic: hidden ruins, secret temples, exotic animals and wild people that carried spears and bows and arrows. Suspicious as I was, given that the jungle was in the middle of Stoke-on-Trent and surrounded by semi-detached houses, I felt it was necessary to be fully prepared.

I packed my little rucksack with all the essentials for such an expedition. It included a hand-drawn map that my grandfather had 'discovered' in the attic, an old military compass, a plastic pirate's cutlass (in case I encountered any fearsome warriors) and some rations in the form of a banana sandwich. And with that, I set off down the garden path with a mission to find the lost treasure. Following each discovery, I became more enraptured with the idea that one day

I might set off on a voyage of my own, to a real jungle. Perhaps even the Amazon rainforest itself.

I finally fulfilled this boyhood dream in 2019. I had spent some time in jungles before, beginning with six weeks sleeping in hammocks and getting bitten by mosquitoes in Belize as part of survival training in the Army. I had walked the length of the Nile and the Himalayas west to east, traversing stretches of tropical rainforest along the way. I had also spent a few weeks in the Congo, exploring the Virunga region with its population of gorillas and forest elephants.

But the Amazon was a different proposition altogether; an area that had haunted my imagination for years. It was the stuff of legends, and I felt a mixture of excitement and trepidation to be finally visiting this iconic landscape. It was a wilderness that had thwarted countless adventurers, inspiring tales of hidden gold and lost cities. As I boarded a fishing boat at the port of Manaus and set sail upstream, I remember being awestruck by a symphony of sounds – the chorus of cicadas and primaeval growls of howler monkeys. I was an interloper in this domain, and felt within me the rainforest's ancient, inscrutable power.

I have been travelling the world for over 20 years now, mainly on foot. This slow method of travel has allowed me to immerse myself fully in the environments that I am treading through and to get eye-to-eye with the people who live there. My journeys have taught me to stay present in the moment and observe the rhythms of life around me. I have seen with my own eyes the tragic effects of deforestation on communities and cultures. How, in my lifetime, we have witnessed the greatest loss of forests since the last ice age. All because of human impact.

The sad reality is that for many people these days, trees are simply resources – things that hold only monetary value as a potential desk, building material or a piece of charcoal to burn. Often that view is held by those who are desperate due to poverty and need to fell a tree to feed their family. At least that is excusable. But for those who willingly forsake nature, either out of ignorance or greed, there can be less sympathy.

I returned to the Amazon a couple of years later on a very specific quest. My mission took me to the state of Acre, a remote region in Brazil, close to the border of Peru. The journey to get there was intense, including six plane rides from London, the last of which was a small, rickety four-seater. As we flew over the forest, I could not take my eyes off the view below me, a thick carpet of vivid greens. At first the forest seemed endless, but then scars would appear on the landscape, like rips in an ancient tapestry. Deforestation had transformed vast areas into grazing and pastureland, the timber being sold off, and great herds of cattle taking the place of the trees.

Since my first visit in 2019, the Amazon may have lost up to 23.7 million hectares of forest, an area almost as large as the entire United Kingdom. It was a shock to the system, a tug of war between man and nature playing out before my eyes.

I had come to the forest to meet Benki Piyãko, a revered spiritual leader of the indigenous Asháninka people of Apiwtxa. He was a shaman, a medicine man famed for visiting the Dalai Lama and inspiring the tribes to unite against corporate greed. Local villagers, under Benki's guidance, had planted over 2 million trees and waged a battle to preserve their land and culture. Their way of life is a recipe for environmental protection and unity with nature. I spent ten days living with the Asháninka people, participating in their ceremonies and helping to replant trees.

When I asked Benki what had inspired him to do his work, he looked at me with serious yet sympathetic eyes. 'It's simple,' he shrugged. 'We are the trees, and they are us. We are not apart from them. We are nature.'

Shortly after I left the Amazon, a combination of devastating droughts, wildfires, and ever-expanding illegal logging resulted in most of Benki's trees being destroyed. In the face of such destruction, it is hard to be optimistic about the future of forests, and yet there is still hope amid the despair.

Throughout history, indigenous cultures like Benki's have harboured a deep reverence for the natural world and have passed down wisdom that recognizes the interconnectedness of all life. From cultures that range from the Amazon to the Congo, intimate knowledge of the land and the indigenous people's harmonious coexistence with nature can provide invaluable lessons for us all. By embracing their wisdom and incorporating it into modern conservation efforts, we can forge a path towards a more sustainable and respectful relationship with the natural world.

Benki's story and message was a revelation for me. It made me look back on my own life and expeditions differently. Perhaps he was right. Despite all the environmental destruction I had witnessed, the answer was in plain sight and has been told over and over again throughout our human history, but we have not always been listening to it. We are more connected to trees than we could ever have imagined. I vowed to discover more and set out on a mission around the world to answer a critical question: does rekindling our ancient connection with trees hold the key to addressing the global crises we face? And what rewards might await us if we succeed?

In this book, I would like to take you on a global journey of discovery. To look at the world through the eyes of our arboreal cousin, the humble tree, and to examine the profound significance of forests. In order to explore our human relationship with trees, I have structured this book into three parts: The Tree of Life, Paradise Lost and Return to Eden. The story of forests is the story of humanity, in all its complexities. Of course, the true, detailed story of trees cannot be condensed into a narrative as simple as this. However, I trust this framework will emphasize a crucial point: that being human means being inherently connected to nature.

In our pursuit of progress, we often forget this bond, leading to the destruction of the very ecosystems that sustain us. By reconnecting with trees and rebuilding these ancient partnerships, we can not only find solutions to many of our current problems, but also rediscover a more harmonious and sustainable way of living that enriches our understanding of ourselves and our place in the natural world.

Few entities have played as vital a role in shaping our entire world and the planet's history as the majestic and enigmatic forests. These sprawling and diverse ecosystems, with their towering trees and intricate web of life, have not only provided us with a home but also influenced the very essence of our existence.

For a very long time, people have looked to trees for knowledge. The origin of the word for Tree in itself is revealing. In Old English it was *Treow* and before that its Germanic ancestor was *Trewa* or *Drewom*.

In Ancient Greek, Tree was *Dendron*. Go back to prehistory and their Indo-European roots intertwine. For the prehistoric nomads of Western Asia, a tree was *Doru*, which in itself derived from *Deru*, meaning to be firm and unwavering, to show commitment or make a promise. Or, in simple terms, to be *True*. It is no coincidence they sound similar, for the tree is truth.

I am no tree expert. I simply happen to have an appropriate surname and an affinity to the wild. But in the process of researching and writing this book, I have spoken to countless people who are. The word 'amateur' is often used critically, to suggest naivety, but the root of the word is Latin for love. If we have no love for nature, it seems unlikely that we will protect it. And without spending time in nature, often and with purpose, it is hard to see how we can nurture that love. But by exploring the interconnectedness of all life, we can uncover the potential for a hopeful future, one where we prioritize the protection of wilderness areas, conserve biodiversity and restore the delicate balance between humanity and nature.

Together, we can ensure the preservation of these vital ecosystems and secure a sustainable legacy for generations to come, and to understand that there is no us and them.

In the words of Benki Piyãko: 'We are the trees, and they are us.'

Levison Wood
London, 2024

Act One

The Tree of Life

Say the planet is born at midnight and it runs for one day. First there is nothing. Two hours are lost to lava and meteors. Life doesn't show up until three or four a.m. Even then, it's just the barest self-copying bits and pieces.

From dawn to late morning – a million million years of branching – nothing more exists than lean and simple cells. Then there is everything. Something wild happens, not long after noon. One kind of simple cell enslaves a couple of others. Nuclei get membranes. Cells evolve organelles. What was once a solo campsite grows into a town.

The day is two-thirds done when animals and plants part ways. And still life is only single cells. Dusk falls before compound life takes hold. Every large living thing is a latecomer, showing up after dark. Nine p.m. brings jellyfish and worms. Later that hour comes the breakout – backbones, cartilage, an explosion of body forms. From one instant to the next, countless new stems and twigs in the spreading crown burst open and run. Plants make it up on land just before ten. Then insects, who instantly take to the air. Moments later, tetrapods crawl up from the tidal muck, carrying around on their skin and in their guts whole worlds of earlier creatures.

By eleven, dinosaurs have shot their bolt, leaving the mammals and birds in charge for an hour. Somewhere in that last sixty minutes, high up in the phylogenetic canopy, life grows aware. Creatures start to speculate. Animals start teaching their children about the past and the future. Animals learn to hold rituals.

Anatomically modern man shows up four seconds before midnight. The first cave paintings appear three seconds later. And in a thousandth of a click of the second hand, life solves the mystery of DNA and starts to map the tree of life itself.

By midnight, most of the globe is converted to grow crops for the care and feeding of one species. And that's when the tree of life becomes something else again. That's when the giant trunk starts to teeter.

– Richard Powers, *The Overstory*

Chapter 1

In the Beginning

———

The clearest way into the Universe is through a forest wilderness.
— John Muir, *John of the Mountains*

Once upon a time there was a man named Methuselah. He was said to have lived before the time of the Great Flood that engulfed the Earth some 5,000 years ago. According to the biblical legend, Methuselah, son of Enoch, was a contemporary of the famous Noah, and lived for an extraordinary 969 years.

Modern sceptics regard this unlikely age as a myth, a translation error, or a symbolic number. Whatever the truth of the matter, Methuselah became synonymous with longevity and gave his name to a life form that did emerge almost 5,000 years ago and survived not for a mere 969 years, but longer – to this day.

Hidden among valleys of the Inyo National Forest of California lies a grove of ancient Bristlecone pine trees. After five hours driving north out of Los Angeles by car, I arrived at the end of the road sometime after three in the afternoon. Just as the sun was retreating behind the White Mountains, casting a long shadow across the mysterious valley, we set off on foot. The pines all around us looked ancient, their bows withered, torn and snakelike. I wondered what calamitous phenomena these majestic beings had witnessed over the millennia. Earthquakes, fires and us, humans, with all our hubris and meddling intent. And yet, here they were, standing proud and triumphant – a symbol of defiance and unlikely survival in the old West.

The trail wound east, following the contours of the ridgelines and offering occasional glimpses towards the aptly named Death Valley some 20 miles distant. Death, it seemed, was never far away in this place. Jagged stumps, torn branches and desiccated tinder littered the forest floor. The scree was loose and dry underfoot from lack of rain. I wondered aloud what wild creatures could lurk in such a ruinous place, but my companion Charles assured me that predators did not dwell at this altitude. He had not even bothered bringing the usual bear spray. The only sign of mammal life was a solitary chipmunk scuttling through the scrub.

As we followed the meandering rocky path around an escarpment, excitement and solemnity competed in symbiosis. After coming from the bustle of LA, the silence of the desert was so conspicuous that it seemed almost loud. No cars, nor even birdsong. Only the crunching of stone beneath our pilgrim footsteps. Our venerated destination was no man-made church or brick-built temple, but one of nature's most sacred offerings. We were in search of a thing so holy that a shaman described it as 'the antenna of the gods'.

It took just over an hour of hiking to find the spot that we had been somewhat unreliably informed was that of this elusive tree. The US Forest Service has deliberately removed any reference to the location of the actual Methuselah tree in order to protect it from the grasping and damaging hands of tourists. There are a few images online, but most of them are of another, more picturesque pine in a nearby valley, so the vast majority of visitors go home thinking they have found the right tree, but in reality have only met Methuselah's far younger cousin.

I say far younger, but the reality is that most of these trees predate the walls of Rome or even the gardens of Babylon. And the truth is that there are probably others older than Methuselah hidden away in some ancient grove unrecorded by the pen of man. As we wandered, deliberating which one of these twisted stumps was the object of our search, I began to think that, in reality, it mattered not which pine was the sire of this grove. They were all connected, regardless of which predated the other. Each and every one, like a large tribe, was related. Some were distant relations, others more immediate offspring, some separated by hundreds of years, others by thousands.

Far from the beaten track, on a steep slope shaded by escarpments of red rock, lies a tree so battered by time that it almost looks dead. Almost. This gnarled and twisted tree is very much alive and has been around for 4,800 years. The age of trees can easily be calculated if they grow in an environment where there are distinct seasons. In summer, when there is plenty of sunshine and moisture, they grow quickly and produce large wood cells; in winter, when growth is slow, the cells are smaller and the wood more dense.[1]

This produces annual rings in the trunk, which is how we know that this tree began its life when mammoths still roamed the Earth, and predates the pyramids of Giza by hundreds of years. It emerged out of the biblical Great Flood, and has outlived the empires of Assyria, Alexander the Great, Rome and Great Britain, to name but a few. This tree has survived every major war in history, and is older than all of our popular religions.

She had been given a man's name, Methuselah, and yet she was most certainly a Goddess. Cool to touch and yet warm in her embrace. Immovable, unshakeable, resolute. Her gaze across that silent valley belied a timeless consciousness and understanding of the seasons, of the land and the stars that we are only now beginning to comprehend.

As the cool air settled among the pines and the sky reddened to a warm winter glow, we sat among Methuselah's roots in a symbolic ritual of thanksgiving. This was the mother tree, to which we owed our very existence. She had been giving us the air we breathe for five millennia. Her roots were intertwined in the deep shale, snarled around rocky outcrops and twisted like the embrace of a serpent or the double helix of our own DNA.

I felt thirsty just looking at her shrivelled and barren trunk. She looked so dry. But deep inside and underground, she had been storing moisture for centuries, slowing her growth and shedding her needles so as to avoid any unnecessary effort. She was merely sleeping, waiting – rain, or snow, would come to replenish her soon. She could wait.

Trees live on a timescale unfathomable to us, and yet they live nonetheless. They breathe, they communicate, they listen and they respond in their own way.

The National Park Service would have you believe that the Methuselah tree is the oldest tree on Earth, but truly ancient trees are not limited to Northern California. The Llangernyw Yew, nestled in an old Welsh churchyard, stakes a 5,000-year claim, as does the Fortingall Yew in Scotland. In Japan, the Jōmon Sugi cedar tree is purported to be 7,000 years old, although since it is hollow we will never know.

If we throw tree clones into the conversation – those which are able to self-replicate from their original roots systems – then we can go back even further. Old Tjiko is a spruce in Sweden whose roots are 10,000 years old. The Juropa oak is a clonal colony tree that may be 13,000 years old. But all of these claims pale when compared with the Pando cluster of quaking Aspen in Utah, which, as a system of identical clone trees, stretches back 80,000 years to the beginning of the last ice age.

The ancestors of these trees were some of the first forms of life on Earth, spectators of the changing of climates and the rise and fall of countless species that developed around them. It is difficult to imagine just how short a time our own species has existed on Earth, but it is something that we should attempt to grapple with so that we can understand how small a role we have had in the grand scheme of time.

Let's go back to a world without trees. Four billion years ago our planet was exactly that: a dystopian Armageddon – the atmosphere filled with carbon dioxide, with violent storms and volcanic eruptions dominating the landscape; a world utterly different from the one we know today. And yet it was a world waiting to give birth to life, a planet pregnant with potential. Despite this tumultuous backdrop, the miracle of life began to unfurl. The first stirrings of existence emerged from the depths of the ancient oceans, as tiny, single-celled organisms wrestled with the challenges of their environment.

The development of life on Earth has been a remarkable feat, achieved against all odds. The environmental conditions that favoured the emergence of early life forms were a complex interplay of chemistry, geology and cosmic happenstance. Deep beneath the surface of the oceans, hydrothermal vents provided the perfect crucible for the formation of organic molecules. These vents, rich in minerals and driven by geological activity, served as the cauldron where amino acids, the building blocks of life, first came into being.

From this 'primordial soup', the first microorganisms arose, simple yet resilient, the pioneers of life on Earth. Over time, utilizing constant energy from the sun, these building blocks allowed proteins to flourish and eventually create microorganisms that multiplied and spread across the globe.

Humans and trees are more closely related than one might think, sharing a common ancestor in the distant past. When eukaryotic cells (organisms whose cells contain a nucleus and organelles) evolved about 2 billion years ago, life on Earth began to diversify. At this point, organisms began to develop different strategies for survival. Over time, one significant evolutionary split occurred, leading to two major branches of multicellular life. One group, the ancestors of modern plants, evolved to harness energy directly from sunlight through the process of photosynthesis, creating their food from light, carbon dioxide and water. These organisms became the green plants we see today, including trees.

The other group, the ancestors of animals, took a different path. Instead of producing their food, they evolved to consume organic matter, either from plants or other organisms. This shift set the stage for the rich diversity of animal life, including humans.

Although this evolutionary split marked a departure from our common ancestry with trees, our genetic code reflects that shared origin. We still share 45 per cent of our genetic material with trees and plants at the molecular level.

If you were to dive into the warm oceans 520 million years ago, you might encounter all sorts of strange sea life – trilobites and other invertebrates hunting, scavenging and filter feeding. Things on land were a different story, though. For starters, the atmosphere had oxygen levels much lower than today, uninhabitable for humans without assistance. The Earth's continents were covered in dry, rocky landscapes, and the primary form of life was mere patchy films of microbes spread across the surface and occasional algae near water.

It is fascinating to think that during the early Cambrian period, as it is known, there was no variety of terrestrial life, let alone anything resembling a tree. And yet, despite that miserable scene, this period played a crucial role in the evolution of all multicellular animals that exist on Earth today.

Fossils from the Cambrian period reveal the appearance of the first arthropods, molluscs and early chordates – fish and animals with a backbone. The emergence of hard shells, vertebrae and exoskeletons allowed these creatures to explore new ecological niches and ushered in an arms race of defence and predation.

Around this time a kind of green algae made its way out of the ocean and clung to coastal rocks, making it the first photosynthetic organism to colonize the land. Over the course of many millions of years it slowly spread its slimy tentacles, changing the terrestrial landscape forever. As the continents merged into the supercontinent Gondwana, plant life started evolving, with simple forms like algae and mosses appearing.

Fast forward another 70 million years and the Earth experienced a biological event of unprecedented magnitude. In the geological blink of an eye, life on our planet erupted into a dazzling array of forms, a radiant burst of diversity that has captivated the curiosity of scientists for centuries. 'The Cambrian Explosion' was not a single, uniform event, but a series of spurts of diversification occurring over tens of millions of years. It was a time of experimentation, with countless species attempting novel body plans, methods of locomotion and feeding strategies. Many of these experiments did not survive the test of time, but some laid the foundations for entire lineages that persist to this day.

The evolution of plants on land is considered one of the major biological innovations of the period. As plants spread across Earth, they absorbed significant amounts of carbon dioxide from the atmosphere. This absorption led to a recycling of carbon dioxide, which in turn resulted in a 'greenhouse effect', causing the climate to become more stable for longer periods. Gone were the extreme, uninhabitable conditions and in their place were more amenable fluctuations in temperature with periods of warming and cooling, and while they triggered several mass extinctions, they also enabled new adaptations and life forms to exist.

During the Devonian Period (around 419 to 359 million years ago), plants such as mosses, liverworts, and eventually more complex, early tree-like plants began to thrive on land. This epoch marked a crucial milestone in the colonization of terrestrial environments. These early land plants adapted to survive in harsh conditions and developed vascular tissues that allowed them to transport water and nutrients more efficiently. As a result, plants grew ever larger and more complex, forming dense forests and providing new habitats for a variety of organisms.

New York is not a place one usually associates with forests, but rewind 386 million years and it was home to one of the world's oldest known forests. Stretching from Hudson Bay to Pennsylvania, it is one of the few places in the world where fossil records reveal the root systems of individual trees within the ancient soil from a time before the dinosaurs. Like dinosaur footprints, these root structures provide scientists with insights into the behaviour of living organisms – and, in this case, the dynamics of forest ecology.

But New York's ancient trees have recently been dethroned. I grew up in England as the son of a geology teacher and, as such, my childhood was spent collecting fossils on the beaches of our little island. I was excited, therefore, when, during the process of writing this book in March 2024, I heard the news that Dr Chris Berry, a British palaeobotanist, had announced a remarkable discovery.

After years of digging around the cliff faces of the Bristol Channel, Berry announced that he and his team of researchers from the universities of Cardiff and Cambridge had unearthed what may be the oldest tree fossils on Earth – just a stone's throw from a Butlin's holiday resort.

Four million years older than the fossil forest discovered in New York, nestled in sandstone cliffs, these fossils reveal a forest of *Calamophyton* trees – prototypes of modern trees. Imagine thin, hollow trunks about 2–4 metres tall, and, more bizarrely, their branches covered with hundreds of twig-like structures instead of leaves. They reproduced not by seeds but through spores that would be carried by the wind for miles. These early trees shed twigs and branches onto the forest floor, creating a nutrient-rich habitat for the first land-dwelling invertebrates.

'When I first saw pictures of the tree trunks,' said Dr Berry, 'I immediately knew what they were, based on 30 years of studying this type of tree worldwide. It was amazing to see them so near to home. It is our first opportunity to look directly at the ecology of this earliest type of forest, to interpret the environment in which Calamophyton trees were growing, and to evaluate their impact on the sedimentary system.'[2]

The region where these fossils were found tells an even broader story. During the Devonian Period, what is now southern England was part of a landmass connected to present-day Belgium and Germany, where similar Devonian fossils have been unearthed. This hints at the widespread extent of these early forests and their impact on shaping the planet's surface.

'The Devonian fundamentally changed life on Earth,' said Berry's co-author, Professor Neil Davies from Cambridge's Department of Earth Sciences. 'There wasn't any undergrowth to speak of and grass hadn't yet appeared, but there were lots of twigs which dropped and had a big impact on the landscape.'[3]

Rivers, which had once flowed freely over barren plains, began depositing sediment that trees like Calamophyton stabilised with their primitive root systems. It was the first time that such tightly packed plants had grown on dry land and their continual disposal of litter shaped the way in which English rivers flowed.[4] 'It changed how water and land interacted with each other,' said Professor Davies. 'The evidence contained in these fossils preserves a key stage in Earth's development, when rivers started to operate in a fundamentally different way than they had before, becoming the great erosive force they are today.'[5] Earth had reached a pivotal moment in which its entire surface would be changed – all because of fallen twigs.

By the late Devonian Period, forests had become lush and diverse as new types of trees evolved. The spread of vegetation helped to stabilize the soil, reduce erosion and create a more hospitable environment for other forms of life. Their ability to absorb carbon dioxide and create stable environments has had profound effects on the evolution of life on our planet. As plants continued to evolve and diversify, they interacted with insects and other animals, fostering complex ecological relationships.

The rise of trees and other plants was crucial in shaping the ecosystems of the time, influencing climate, and enabling the evolution of diverse life forms. So that by the time we reached the Carboniferous Period (around 359 to 299 million years ago), the humid climate with its vast swamp forests gave rise to ferns, conifers and a smorgasbord of other plants and trees.

Compared to the cacophonous noise of a modern rainforest, or even British woodlands, these ancient Devonian and Carboniferous forests would have been eerily silent. No herbivores yet existed that could eat the new plants, leaving them largely unmolested to grow in peace.[6] Nevertheless, the falling detritus created a planet covered in soil, which now provided a warm, moist home for a variety of millipedes and an array of insects. It was from these early bugs that vast dragonflies, some two feet long, and giant spiders emerged.

Following them came huge amphibians to eat the insects, and then reptiles – the forerunners of dinosaurs. At this point, as the

Carboniferous gave way to the Permian, the Synapsids also emerged, introducing to the world a very strange creature, half-lizard and half-mammal; an evolutionary experiment that 250 million years later led to the primates that gave birth to you and me.

And all of that because of trees like the ancient *Calamophyton*.

The great green giants had arrived. Life on Earth – as we know it – was born from the conditions that these amazing organisms created. From the moment we split in the evolutionary tree, they have been fostering our environment for shared benefit and continued survival. We owe them everything and it is time we started to treat them with the respect they deserve – as living, breathing entities.

Chapter 2

Treespiracy

———

It is not the strongest of the species that survive, nor the most intelligent,
but the one most responsive to change.

– attributed to Charles Darwin

'The trees are alive!' scream the hobbits, as the giants emerge from their slumber to join the fight against the forces of evil. They are led by Treebeard, oldest of the Ents – ancient, mythical beings who dwell in Fangorn Forest. In Tolkien's fantasy it is the Ents, wise and deliberate, who are the guardians of the forest and the shepherds of the trees.

Back when the Oxford don published *The Lord of The Rings* in 1954, little was known about how trees communicate with each other. It was not until several decades later, in the late 20th and early 21st century, that researchers began to uncover the complex ways in which trees and plants 'talk' and how this might demonstrate a sort of intelligence.

Now, modern science is revealing that Tolkien was remarkably accurate with his description of Ents and their language; that their communication is incredibly slow by human standards. In the world of trees, everything is in slow motion. Their long lives give them a different timeline to us. While we measure our lives in decades, trees measure theirs in centuries and sometimes millennia.

Over a hundred years ago, the renowned biologist Charles Darwin proposed the intriguing idea that plants possess a brain-like structure in their root tips. It was a heretical suggestion, like so much of his work, and even though he was somewhat off the mark with this one, modern research does indeed show that while plants do not have brains, the way they act communally suggests a form of intelligence.

We already know that plants turn their leaves towards the light, which in itself is an ability to sense, but more and more scientists are in agreement that plants have other senses too and can not only detect sunlight and temperature, but also sniff out chemicals, feel vibrations and even register music. What's more, they even appear to be able to communicate with each other.

Scientists are beginning to understand that plants are much more 'intelligent' than we have ever given them credit. In our short stint on planet Earth, self-absorbed humans have often regarded plants and trees as merely a staging post for our theatrics; a backdrop to the fast-paced dramas, relationships, tragedies and triumphs of our own existence. In the short time we have been here, we have waged wars over vast lands, altered the entire climate system, built empires and cities. In 'developed' societies, we are so obsessed with our own intelligence and innovation that we have often forgotten the knowledge and teachings from the trees that have lived on this Earth for so long before us.

While many people are fascinated with the sci-fi notion that we might one day meet and communicate with aliens from far-away planets, it is only recently that we have remembered that it might be worth talking to the giants that live on our own planet. It has not always been this way; early societies respected nature, and indigenous communities around the world still heed its wisdom.

Indigenous peoples have traditionally observed and interpreted natural signals, known as bioindicators, to guide their activity and survival. This knowledge has been passed down through generations and remains relevant today.

My journey had taken me deep into the Arctic Circle, to Northern Finland. I had come here to learn from the Sami, an indigenous people whose lives are intertwined with this frozen wilderness. In these remote regions, spanning across northern Norway, Sweden, Finland, and the Kola Peninsula in Russia, the Sami have thrived for centuries, guided by a profound understanding of the natural world. I trekked across the crisp, white plains with my Sami guide, Maria. I marvelled at her ability to read the landscape like a map. She explained that the Sami have long relied on these natural signals to guide their activities and ensure their survival in these harsh conditions.

As we paused near a small grove of birch trees, Maria gently touched one of the trees and pointed to the ground. 'The birch trees tell us a lot about what's coming,' she said, her breath forming wispy clouds in the cold air. I watched as she examined the fallen leaves. 'If they drop early, we know a harsh winter is ahead.'

Reindeer herding, a cornerstone of Sami culture, relies heavily on understanding the subtle changes in snow and ice conditions. Maria showed me how the reindeer's behaviour can predict weather patterns and seasonal changes. Studies have also shown that the Sami's knowledge of their local conditions, gleaned from generations of observation, often aligns with modern meteorological data. Their deep, intuitive connection to the world around them gives the Sami people a unique ability to read nature's signals, an invaluable skill in a world that often forgets to listen. 'Nature is so knowledgeable, we don't need technology to learn from it,' Maria told me. 'Nature is intelligence.'

Our modern societies are un-superstitious and grounded in science, but if we are willing to think beyond this, towards alternative non-materialistic avenues of perception, we might just discover the true consciousness of nature. Ecologist and philosopher David Abram describes it as nothing less than 'magic'. He writes:

> In its perhaps most primordial sense, it is the experience of existing in a world made up of multiple intelligences, the intuition that every form one perceives – from the swallow swooping overhead to the fly on a blade of grass, and indeed the blade of grass itself – is an experiencing form, an entity with its own predilections and sensations, albeit sensations that are very different from our own . . .[1]

Abram's perspective comes from visiting native cultures around the world to learn from their understanding of mankind's connection with nature and combining this with scientific knowledge and a philosophical approach to humans and nature interacting. In his book *The Spell of the Sensuous*, Abram discusses the idea that the natural world, including trees, has its own forms of sentience and subjectivity. He believes that the environment communicates with us in various ways.

While still a subject of much debate, there have been a lot of studies suggesting that plants might be able to make sentient, intellectual decisions. Proponents of plant neurobiology suggest that rather than simply being hardwired to respond to certain external stimuli in predetermined ways, plants may exhibit more complex and purposeful behaviour.

Researchers at the University of Murcia placed potted French bean plants in cylindrical booths, with some plants placed alone and others with a cane planted 30 centimetres away. They found that the shoots of the bean plants grew along more predictable paths in the presence of the canes, as though they had an awareness of their presence nearby.[2]

'It is one thing to react to a stimulus, such as light, it is another thing to perceive an object,' said one of the researchers. 'If the movement of plants is controlled and affected by objects in their vicinity, then we are talking about more complex behaviours, not reactions, and we should be able to identify similar cognitive signatures to those we observe in humans and some animals.'

Consciousness extends beyond the human experience. It is an intricate mental state, encompassing awareness, perception, and subjective experiences. In the world of animals and other non-human life forms, consciousness is shown by their ability to see, hear, feel and interact with what is around them. This shapes their individual worlds. From the vibrant coral reefs teeming with life to the instinctual wisdom of ancient trees, non-human consciousness manifests in various forms.

While trees lack nervous systems and brains akin to animals, their complex responses to environmental stimuli challenge traditional notions of sentience. They have a remarkable ability to sense their environment, communicate with other trees through chemical and electrical signals, differentiate between kin and non-kin, and respond to external stimuli such as light and touch, with behaviours that enhance their survival. This remarkable adaptability has evolved over millions of years, and we are only beginning to grasp its full extent.

In pursuit of a deeper understanding of this extraordinary aspect of tree life, I found myself venturing into one of the world's most renowned rainforests. I set off from the isolated river-port city of Manaus up the Rio Negro, a tributary of the Amazon, in search of a legendary tree. Of the more than 80,000 plant species to whom the forests of the Amazon basin are home, perhaps the most fascinating is *Socratea exorrhiza*, or as they say in Portuguese, the *Palmeira ambulante*.

My guide, Diego, a native of the river people, led me aboard a small wooden boat. We motored our way up river for two or three hours past sporadic indigenous villages where children and dogs ran playfully through the cultivated banana fields. Women, with colourful dresses and painted faces, watched from the banks and a man, presumably the chief, with his feathered headdress and a necklace of jaguar claws, stood in silence.

'They are the Desana tribe,' said Diego. 'The people of the anaconda. They believe they came from the Milky Way, the celestial snake, and their task here on Earth is to maintain the natural order. They fear the wrath of the energy of the sun if harmony is out of balance. The master of the animals will reap his revenge and the spirits of the water will come to take any man who thinks himself above the level of the forest. It means they never hunt more than they need.'

Leaving behind the seemingly last remnants of human habitation, our canoe sped further upriver until finally we came to a stop by a muddy bank. Diego helped me off and we clambered up a slippery path into the forest. Fat rubber trees and towering tucumã palms, with their rings of wicked thorns, seemed to be spaced out with little ground vegetation.

It was unlike the dense, scrubby jungles I had encountered before in Central America, which were mostly secondary forest, the regrown vestiges of former plantations or areas that had been previously logged in the last century. Here, walking was not difficult. The tall canopy allowed so little light through that the undergrowth was forgiving and natural paths weaved between the high trees.

Diego pointed out the tracks of a peccary, a small pig-like creature that provided easy prey to the king of the South American jungle: the jaguar. We walked on through the forest, keeping a wary eye in case we spotted the elusive beast, but apart from the occasional howl of a monkey and shrill of cicadas, there was no other sign of mobile life. That was, of course, until we reached what we had come to find. 'There,' Diego pointed, *'Palmeira ambulante.'*

The 'walking palm' has developed a very particular method to ensure that it can thrive in the rainforest – and the clue is in its name.

Tall, thick roots protruding up metres from the ground greet the uninformed traveller, who has to strain to see the massive underbelly of the tree growing among the branches of others. These stilt-like roots extend, insect-like, across the ground, creating the effect of a tree caught striding across the rainforest floor using its roots as legs. Upon first coming across these curious trees, scientists believed that the roots had worked as a sort of buttress, stabilizing the trees; like the thick, wide roots of the kapok tree, which spread out, partially above ground, from its massive trunk. They soon began to realize that this understanding was not accurate.

Rather than working to *stabilize* the tree, the walking palm's roots offer it new life. When the palm is sapped of the sun after becoming overshadowed by neighbouring trees, or starved by a soil that has become stripped of nutrients, the tree gradually produces roots that stretch into a new direction, very slowly relocating to a more beneficial location from which to sup.

Whether or not this constitutes 'walking' in a true sense remains a subject for debate, but for the indigenous communities that inhabit the rainforest, they are in no doubt.

'It walks anywhere between two and twenty metres a year,' said my guide Diego. 'Always looking for the best soil. It's a very clever tree.' Clever indeed. The tree's fruits are edible, and its roots have been utilized as both an aphrodisiac and a medicine for its effects: it is said that they help to heal hepatitis.

This tree, clever as it is, dwells on the edge of a forest whose depths have yet to be plumbed by all of scientific curiosity. We have here but one wonder of a myriad that lies concealed beneath the Amazonian canopy. It would be bold to suppose that we have already seen all that trees have become in their long, slow development. What other secrets this verdant labyrinth still holds, only time will reveal.

If we take anthropologist Jeremy Narby's definition of intelligence as 'adaptively variable behaviour within the lifetime of the individual' – basically, the ability to learn and change – it is fair to say that trees do exactly that.[3] These debates underscore the need for a deeper understanding of the intricate relationships that trees and forests have with their ecosystems and with one another and the nuances of tree intelligence. Perhaps trees are more similar to us than we previously thought.

There is a famous philosophical thought experiment that asks: 'If a tree falls in a forest, and there's no one around to hear it, does it make a sound?' It is a question that exploits the tension between perception and reality, but it also highlights the blinkered nature of the modern human condition. Some argue that, if we define sound as our perception of air vibrations and there is no ear to pick them up, then there is no sound. But in our typical self-centred nature, we often fail to consider the fact that trees can *also* perceive sound vibrations. They might not have ears, but they hear nonetheless. And in a form as ancient and enduring as they are, imagine if they could talk – what a story they could tell.

The poet Hermann Hesse once said, 'Whoever could know how to listen to trees could discover the truth of the world.'

Perhaps we have been too quick to dismiss our arboreal cousins. More and more research shows that trees communicate with each other, warning others of danger and protecting their communities in forests. These gentle giants, while seemingly quiet and inconspicuous, have been talking to one another long before we walked this Earth. While we do not know if plants can feel pain, they certainly do not like being eaten. Many plants, when being munched on by an insect – a caterpillar, for example – will release defence chemicals to make them less tasty, or even to attract types of birds that will do their job for them and eat the insects. They will also warn other plants around them to do the same by releasing an airborne chemical signal.

In 1997, Dr Suzanne Simard stunned the world of science with her research, which showed that forest trees use fungi to communicate with one another. In her influential work *Finding the Mother Tree*, Simard 'set out to figure out where we had gone so very wrong, and to unlock the mysteries of why the land mended itself when left to its own devices.' The Canadian scientist describes how 'the trees soon revealed startling secrets'.[4]

In a symbiotic relationship, the trees and the fungi help each other to survive and thrive. Interestingly, from a taxonomic perspective, fungi are more closely related to humans than to plants.

'It always amuses me to ponder how the mushrooms thrust into the vegetable section in supermarkets must be silently spitting with rage at their misplacing,' wrote Merlin Hanbury-Tenison in *Our Oaken Bones*.[5] Perhaps we should take heed of these relationships. These linkages with fungi enable trees to take care of their own offspring, pass nutrients to family members and share information about the environment – the 'Wood Wide Web', as Simard describes it.

She discovered these exchanges by isotopically labelling carbon and watching it move from one tree to another via the fungal networks. 'All trees are in a web of interdependence,' wrote Simard, 'linked by a system of underground channels where they perceive and connect and relate with an ancient intricacy and wisdom that can no longer be denied.'

Picture this, a magical realm of abundant life, thick tree roots intertwined with their neighbours, reaching down deep into the earth. Fungal threads (*hyphae*) wrap themselves around, or even penetrate the roots of all trees, clinging to the tiny hairs and forming a symbiotic partnership known as mycorrhizae. Up to half of a forest's biomass lies hidden beneath the ground in the soil. A mere handful of soil teems with more diverse life forms than there are people on the planet.

This is especially evident in old-growth rainforests. A single gram of soil can contain 90 metres of mycorrhizal mycelium, the stringy network that connects the entire forest. If we were to take the length of mycorrhizal mycelium just found in the top ten centimetres of soil on Earth, its length could stretch halfway across the galaxy.[6]

Simard conducted hundreds of experiments over many years demonstrating the purpose of these mycorrhizal networks and their symbiotic relationships with trees. What happens when a tree encounters a threat such as browsing herbivores or invasive pests? How does it warn other trees? She found that it can send electrical signals through its roots and the mycorrhizal network to neighbouring trees. These trees, upon receiving the signals, can activate their own defence mechanisms, such as releasing chemical compounds to deter herbivores or insects. It is not a fast process, though. Electrical impulses in tree roots move at a rate of one third of an inch per second.[7]

Edward Farmer from the University of Lausanne in Switzerland has studied these electrical pulses, revealing a signalling system based on voltage that bears intriguing similarities to the nervous systems found in animals. This warning is not entirely selfless: by initiating this early warning system, the tree is protecting the entire ecosystem upon which it relies.

In another study, scientists discovered that underground mycorrhizal networks connected healthy plants and pathogen-infected tomato plants. When the healthy plants came into contact with the infected plants, they began to develop and share immunity and develop resistance. In short, they were given not only a heads-up about future attack, but the weapons to fight it with.[8]

But it is not only through the soil that trees and plants pass messages to each other. Just as someone might splash themselves with perfume or aftershave in order to communicate nonverbally with those around them, trees also emit their own form of airborne signals: chemical pheromones. When a tree notices a change in its environment, it releases molecular messages into the air that are transported by the wind to its neighbours. Studies have shown that acacia trees release these volatile organic chemicals (VOCs) into the air when under attack from herbivores such as camels or deer. When nearby trees come into contact with these chemicals, they produce higher levels of defensive compounds in order to prepare for the potential threat.

Experiments with wild tobacco, too, have shown that when the plant's leaves are being eaten by hawkmoth caterpillars, the plant emits odoriferous compounds that attract *Geocoris pallens*, an insect that eats these moth's eggs. Apparently a chemical contained in the caterpillar's saliva triggers the SOS emission. It is not just the affected plants that respond, but others nearby too. It is not a 'one-size-fits-all' message. In response to herbivore attacks, some plants release volatile organic compounds that attract predators of the herbivores, effectively enlisting natural allies for defence.

Whether this communication is intentional or not remains to be seen; perhaps the trees are screaming when under threat, and the other trees simply notice and then react accordingly. It is not a stretch to imagine trees, who have existed through great swathes of history, engaging in a bit of general conversation every now and then. Or perhaps it is the fungi who are pulling the strings after all? The truth is, we do not know for certain. What we do know though is that there is more to the forest than meets the eye.

In evolutionary terms, it makes sense that a plant facing these threats could signal its distress to other plants, so that they could avoid the dangers or better defend themselves against threats. As well as their capability to communicate with their neighbours, trees have the ability to learn, and pass this information on to future generations.

Peter Wohlleben studied the ancient beech forests of Germany for decades and believes that these trees pass on genetic information through their seeds, which increases their chances of survival. Wohlleben studied weather patterns, including spikes in temperature, and discovered that in times of drought, some trees would 'panic' and shed their leaves. This is done in an effort to avoid losing water through transpiration, but comes with its own downside. Shed too soon and you lose the ability to transpire at all until there's new growth. This can result in cells bursting and photosynthesis halting.

It is a dangerous game for a tree to play. If a tree is to be able to survive its mistakes, then it must change its behaviour accordingly to avoid repeating them in the future. And apparently this is exactly what happens. As Wohlleben has shown, these changes can be studied and appear to be localized, suggesting that this ability to learn from experience is no different to how we ourselves learn: either from our own experiences, or from the lessons communicated to us by our family and community. And sometimes, from a stranger.

Trees also communicate with *us*.

In *The Overstory,* Richard Powers wrote that the chemicals released by the roots and leaves of trees affected our mood: 'When you feel good after a walk in the woods, it may be that certain species are bribing you. So many wonder drugs have come from trees, and we have not yet scratched the surface of the offerings. Trees have long been trying to reach us. But they speak on frequencies too low for people to hear.'[9]

We can also be on the receiving end of their distress signals. Many people are drawn to the smell of freshly cut grass. It evokes memories of childhood and picnics, but the reality is much more distressing. Without the ability to scream, grass uses this scent that we all love so much as a sort of aromatic SOS. The smell is a form of communication that tells other plants around it that it is in a state of stress, alerting them of the threat. So next time you get a heavenly whiff of freshly cut grass, I am sorry to remind you that it is grass screaming. Trees do this too when they are cut down or damaged.

In *The Mountains of California,* mountaineer and conservationist John Muir wrote of his sensitivity to the cry of the natural world: 'I could distinctly hear the varying tones of individual trees – spruce, and fir, and pine, and leafless oak – and even the infinitely gentle rustle of the withered grasses at my feet. Each was expressing itself in its own way – singing its own song and making its own peculiar gestures – manifesting a richness of variety to be found in no other forest I have yet seen.'[10]

Muir's sensitivity to these 'songs' of trees underscores the notion that trees possess a form of communication that is rich and varied, much like the complex nature of human language. Recognizing the sophisticated ways in which trees interact with their environment transforms our approach to nature from one of exploitation to one of respect and collaboration; there is much to learn from listening to the intelligent networks that sustain our planet. Paying more attention to them could enhance our lives in more ways than we can imagine.

The British monarch agrees. King Charles III thinks it 'absolutely crucial' to 'happily talk to the plants and the trees and listen to them.' Sometimes he gives a leafy branch 'a friendly shake to wish it well.'

While many have laughed at this attitude, it is intriguing to think of trees 'talking' to us. When we envision the act, it seems a mouth and ears would be required. But, even as humans, much of how we talk to each other is done without words; the nod of a head, the touch of an arm, the crinkle of eyes, even the smell of the food we have lovingly cooked for our partner. If we think of talk from this broader viewpoint, as simply a connection with another being that relays a message, we can begin to understand tree language. For them, it is a vocabulary of chemistry, spoken softly in the soil, and whispers in the wind.

In 2008, the Royal Horticultural Society did a bizarre experiment where they played the voices of ten people to tomato plants through headphones to see which voice they liked the best. The winner was none other than Charles Darwin's great-great-granddaughter, who read from the famous botanist's *On the Origin of Species*. Ms Darwin's reading saw her plant grow 1.6cm higher than the control plant. The more open-minded members of society have long insisted that plants and trees that are sung or spoken to positively will grow better. Could this really be true?

It appears that plants, including trees, respond to music too. In 1962, Dr T C Singh, Head of Botany at Annamalai University, conducted one of the earliest studies of the effect that music had on plants. It was found that the growth rate of balsam plants exposed to classical music increased by 20 per cent, while their biomass increased by 72 per cent.[11] Playing *raga* – a form of Indian classical music – resulted in 25–60 per cent greater growth than the average plant.

The researchers began to experiment with a range of different genres and musical instruments, from traditional Indian dance to the sounds of folkish flute-playing or the bombastic melody of a harmonium. They came to the conclusion that classical music was the most effective genre for plant growth and that the violin was the most beneficial instrument.

Around the same time, a Canadian by the name of Eugene Canby independently replicated these results. By exposing his wheat fields to Bach's violin sonatas, he found that yields increased by two-thirds.

In 2017, a Hungarian study showed that seeds exposed to music germinated more rapidly than those not. 'It is possible,' the authors believed, 'that the music stimulated in some way the plants' embryos, motivating them to germinate.'[12]

One academic determined that the sonic frequencies plants best responded to were best served by Bach, Haydn, Beethoven, Brahms and Schubert, played intermittently for several hours throughout the day. Any more, and the plants began to crumble under the 'weight' of too much music.[13] She also found that apparently plants really do not like heavy rock music. Like humans, plants can be moved by music – and tortured by it!

If trees can talk to each other, appreciate music, and scream for help, then it follows that they could have feelings. But how can a tree feel, you might ask? As we have discovered, trees can react to sounds, pass on chemical signals, and warn others of a threat. The logical conclusion is that trees somehow know the difference between pleasure and pain. Trees 'enjoy' the radiating energy of the sun, and they 'enjoy' classical music. They dislike – for want of a better word – being eaten by beetles, being burned and seemingly, death metal music.

They might not bleed like you and me, but they do have a skin, in the form of bark, which keeps them protected. Trees help each other out and interact with other species. Some, as we found out in the Amazon, might even be able to walk in search of deeper soil, more sunlight and a better life. Just because they do not have a brain, does that mean they cannot feel?

Jellyfish, sea anemones and urchins do not have brains, but they certainly react to their environment and make decisions, just like trees. Because trees do not move (very far) does not mean they are incapable of sensation. Corals, barnacles, and oysters do not move either. But they too have a desire for life. They also can communicate, can transmit, can *grow*. How then is a tree's existence any less significant than that of an animal?

Under a microscope, the needle of a fir tree appears to show a dozen silver lines, made up of tiny dots against a green background. These are breathing pores, each one made from the gap between two curved cells. The cells integrate information about the state of the needle's environment and then open or close the pores to admit gases or release water vapour. Every cell inside the needle is making similar assessments, sending and receiving signals, making decisions and modulating its behaviour as it learns about its environment.

When electrical signals travel through animal nerves, they contribute to processes we refer to as thoughts. Interestingly, the proteins involved in generating these electrical signals in animal nerves are somewhat like those found in plant cells that also facilitate electrical excitation. While the specific proteins and mechanisms differ, both animals and plants utilize electrical signals to respond to their environments.

The fact that these actions take a minute or more to travel the length of the leaf suggests that similar processes are happening, only in slow time. 'Trees have long thoughts, they are long-breathing and restful, just as they have longer lives than ours. They are wiser than we are . . .,' wrote Hermann Hesse. Plants do not have brains, but perhaps their thinking is diffuse, like that of an octopus, where 'thought' is located in the connections of every cell.

Trees are more like us than we ever thought previously possible. Scientists from Hungary, Finland and Austria did a study observing birch trees and were astonished to find that they show signs of a form of rest at night. Using lasers on calm nights, they found birch tree branches lowering up to four inches and then returning to their normal day-time position at sunrise. They extended this study to 22 trees of different species, and identified a repeating pattern every three to four hours of changing branch positions. They suggested that trees might have a potential slow water-pumping mechanism, similar to a heartbeat.

Perhaps trees, in their static, slow-living existence, with their sessile grace, are simply a more refined species. While we humans, the descendants of those frantic little amoeba in the primordial soup, rush around trying to survive and now, in our hubris, having gained the power of language and complex thought, do what with it? We overthink, we try to control and manipulate everything, and in doing so cause untold damage. Trees, however, simply exist in a harmonious embrace with Mother Earth. And what could be more perfect than that? Trees don't lie, they don't steal, they don't kill, or waste their time on fear or greed. They simply live in peace, for a very long time.

Recent research has given us a new understanding of the mysteries of trees, and yet many secrets about their communication methods remain. We live in an exciting time when we might just be on the brink of being able to truly understand nature and find a way to communicate with them. As King Charles has been known to have said, plants respond when we talk to them.

In the 375 million years that trees have lived on Earth, they have undoubtedly learned a great many things. Long before humans, mammals or even the dinosaurs were around, trees thrived on every continent – regulating their environment, manipulating the climate and setting the conditions for other species to join them in the magic of existence. Theirs is a tale of survival, adaptation, and symbiosis and it was the wisdom of these living organisms that enabled animal life to take hold and flourish.

Chapter 3

Mother Earth

———

All things share the same breath – the beast, the tree, the man, the air shares its spirit with all the life it supports.

— Chief Seattle, during a speech

Just as a single life is a tale of triumphs and tragedies, the history of life on Earth is no different. Over the course of the last few hundred million years, our planet has witnessed periods of great evolution, as well as cataclysmic extinctions on a massive scale. The climate has seen times of stability as well as tumultuous destruction, each having their impact on the way that life has been able to adapt. Volcanic eruptions, cosmic interjections, alternate periods of great cooling and heating have all left their mark.

I am sure you already know of the famous asteroid that wiped the dinosaurs off the face of the Earth, but 252 million years before that, an even more devastating series of catastrophic events occurred. The 'Permian extinction' or 'the great dying' resulted in the obliteration of 95 per cent of marine species and three-quarters of land species on Earth. And as if that was not bad enough, a mysterious event occurred that killed nearly all trees as well. It could have been the end of life on Earth.

So what happened? For palaeobiologist Cindy Looy, investigating this event was like 'working on the greatest murder mystery of all time'. Doug Erwin, a palaeontologist at the Smithsonian Institution, agrees: 'It's not easy to kill so many species. It had to be something catastrophic.' Various theories have been presented, similar to those proposed for other mass extinctions – eruptions, impacts and global warming may have played a part, but it could also have been assisted by something else.

Deep in the Dolomites, the Butterloch gorge offers a silent, ancient twist for the extinction hunters. Those who chip away layers of rock in the gorge can discover microscopic fossils of primaeval plants. The layers that are most deeply hidden reveal rich levels of pollen, indicative of the good health of the forest prior to the unknown cataclysm. The layers that developed afterwards, however, do not contain pollen – they contain fungi. Masses upon masses of fungi. The rotting biomass of the Earth's trees were practically *devoured* by this spore.[1] Were killer mushrooms the culprit?

It is difficult to kill 96 per cent of the world's species. Did the murderer descend from the skies or arise from the sea? Whatever the cause, fungal spores flourished as scavengers of the apocalypse, feeding upon the corpse of the Earth's trees *en masse* as the extinction spread across the planet.

Without exoskeletons or internal bones, plant fossils are rare, and biologists have much greater difficulty in determining their evolution. Nevertheless, some remain.

Three thousand miles away, near the Dead Sea in Jordan, palaeontologists came across some fossil plants so well preserved that they could use a small amount of acid to remove rock from the fossil while retaining the plant's waxy sheath.[2] Conifers that had lived and grown 252 million years ago, during this cataclysmic extinction of life, belong to a family of trees still alive today.

While some plants survived the extinction crisis, the nature of the event – and the terrible, parasitic fungi that followed – severely shaped the type of life that came after. The few species that lived on would not only survive but come to *dominate* the world's flora: out of 400,000 plant species alive today, around 300,000 are flowering plants.[3] In more protected areas of the planet, ferns dominated the land. In areas most exposed, flowering plants thrived. Trees, with their long, large roots, fungal dissemination and reliance on the sunlight suffered most.

The near-decimation of the Earth's trees was thankfully not the end. Their renaissance sustained new life on a ravaged planet and that life began to *thrive*. A new Garden of Eden was created in which emerging species could exist – and flourish – in balance. Death had created new opportunities for life; gradually, the thick carbon dioxide that covered the planet was converted into life-giving oxygen.

There have been five mass extinction events in the Earth's history. The first was brought about through extreme changes in sea levels, the second through a severe and rapid global cooling. The third mass extinction was likely caused by global warming brought about through volcanic eruptions that affected land and sea, as was the fourth. The fifth is the one we are most familiar with.

About 65 million years ago, an event occurred that shaped the course of life on Earth. Somewhere off the coast of the Yucatán Peninsula in modern-day Mexico, a giant asteroid crashed into the Gulf sending shockwaves around the planet. The aftermath had catastrophic consequences. The impact covered Earth in a thick cloud of ash that blocked sunlight and triggered a dramatic drop in temperature. Dinosaurs, which had lorded over the Earth for 165 million years, were wiped out in the blink of an eye, their cold-blooded bodies unable to cope with the sudden change in temperature. Only the tiniest lizards, sea creatures and earth-burrowing animals survived; those able to stay warm and protected.

Over the next few million years, it was these small animals, the ancestors of the birds and mammals, that came to take over the evolutionary tree. It is a cautionary tale of how fragile life on Earth can be, and also how interconnected the ecosystem is on Earth. Guess who the constant survivors were? Trees and plants that had evolved to be far more resilient and adaptable to climatic change.

Their deep-rooted connection to the soil and the ecosystem helped them endure the aftermath. Trees had evolved mechanisms such as dormancy, allowing seeds to remain viable through prolonged adverse conditions such as the extended cold and darkness following the impact. This adaptation allowed them to maintain minimal metabolic activity and survive until conditions improved. As sunlight gradually returned and the environment stabilized, these trees resumed photosynthesis, replenishing atmospheric oxygen, stabilizing ecosystems, and contributing to the gradual warming of the planet.

Trees allowed new life on Earth to develop and flourish. Their resilience and ecological functions were key to Earth's recovery. But it was the way in which trees have evolved to work together with other flora, fauna, and the complex ecosystems on Earth, that their success originated from.

The word 'conspire' originally means 'to breathe together', derived from the Latin *conspirare*. Trees, in their efforts to photosynthesize and grow, absorb carbon dioxide from the atmosphere, storing it in their mass. They literally breathe it in and then expire excess oxygen, which animals inhale, of course, expelling CO_2 in return. Animals and trees have evolved to support each other.

Every single animal on Earth – the birds, the mammals, and fishes – contribute, and all the plants return the favour. Phytoplankton, the microscopic algae that live in the sea, produce more than half the oxygen in the atmosphere. Forests play their part in that, too, in producing perhaps 30 per cent of our oxygen, but their real practical value is not in what they breathe out, it is in what they breathe in.

Trees are not merely passive beings, but the facilitators of life itself, wielding the power to manipulate the sun's energy, which transforms every element of creation.

Trees created a home in which they and we can prosper. They congregate in huge numbers, moulding their local environment. They are communal entities, sometimes to the point of being communistic; they can link their foliage to create a closed canopy, or where required, avoid each other to allow in more light. They function for the benefit of the whole and make surprising partnerships with other species.

Trees have relationships and work together above and below ground, communicating and cooperating all the while, acting like a superorganism and controlling everything from the ground up. They generate soil, regulate the temperature and even control the weather across the entire planet.

At the heart of this transformative power lies photosynthesis, the process by which trees convert sunlight into chemical energy. This remarkable feat not only fuels the tree's growth and survival, but also has far-reaching consequences for the entire planet and has fed the evolution of life itself.

In his notebooks, Leonardo da Vinci wrote that 'the sun gives spirit and life to plants, and the earth nourishes them with moisture.' He was right, of course. To understand more about how plants came to shape our world, we must first understand the process of photosynthesis.

Ancient microorganisms possessed a unique ability to convert sunlight, water and carbon dioxide into energy-rich organic compounds, while releasing oxygen as a by-product. This process would become a transformative event in the history of life. The oxygen released by cyanobacteria (a particular group of microbes able to generate energy from sunlight) gradually accumulated in the atmosphere, leading to the oxygenation of Earth's environment.

Over time, oxygenic photosynthesis became more efficient and diverse. The refinement of photosynthesis occurred through a series of adaptations to optimize energy production. As Earth's atmosphere became increasingly oxygen-rich, organisms evolved to use chlorophyll, a pigment that is more efficient in capturing light for photosynthesis. This adaptation allowed for higher energy yields and facilitated the expansion of photosynthetic organisms across various environments.

Soon, photosynthetic organisms developed various mechanisms to adapt to changing environmental conditions. Some plants evolved specialized structures like stomata, tiny openings on their leaves, to regulate gas exchange and minimize water loss during photosynthesis. The oxygen produced by photosynthetic organisms allowed aerobic life forms, including animals and humans, to emerge and thrive. Additionally, by absorbing carbon dioxide, trees acted as carbon

sinks, storing large amounts of carbon in their trunks, branches an
d roots.

Through these processes, trees have played a crucial role in
regulating the Earth's climate and atmosphere throughout history.
What many people fail to realize is the far-reaching impact forests can
also have on global climate systems and weather.

Fred Sacks of the University of British Columbia believed that
'life on Earth depends in no small part on the stomata.' We have
already discussed the significance of carbon sequestration above,
but the fact that leaves both absorb and dispel water is of no small
importance.

As Fred Pearce describes in *A Trillion Trees*:

> A single leaf may carry more than one million stomata, a single
> tree may have hundreds of thousands of leaves, therefore
> hundreds of billions of stomata. A big rainforest like the Amazon
> can have hundreds of billions of trees. That makes a lot of stomata.
> [. . .] They act as valves regulating the inflow of carbon dioxide
> and outflow of water in ways that allow plants to optimize the use
> of both. They can shut down during droughts, for instance. While at
> other times they stimulate the tree to pump more moisture
> from the soil to boost photosynthesis. By regulating this flow of
> oxygen and water for the tree, they also regulate them for the
> planet.[4]

For many years, we have understood that most of the rain that falls
down upon us begins its life as the evaporated moisture from seas and
oceans. Children spend their school years studying the life cycle of
water, committing this to memory. Moisture from the oceans does

indeed get evaporated into clouds, but this moisture usually disappears a few hundred kilometres away from the coast.

Fred Pearce reveals the magical source of inland moisture: trees. Forests, he argues, transpire moisture back into the dry air. 'The result,' he argues, 'is the difference between West Africa and the Amazon.' Trees *create their own rainfall*.

Peter Wohlleben agrees that 'large forests change everything. They suck moist air into the interior of continents – and they do so with such force [like] biotic pumps. Even thousands of miles from the ocean, there is no discernible reduction in rainfall over large, naturally growing forests.'[5]

Through transpiration, trees release water vapour into the atmosphere, thus contributing to the moisture content of the air. All the three trillion trees on the planet release an estimated 60,000 cubic kilometres of water a year: moisture that causes at least half of all rain and snow that falls on land.

In regions with extensive forest cover, the collective transpiration of trees can even create microclimates, altering temperatures and precipitation within their vicinity. It seems counterintuitive for them to pump all of this moisture into the atmosphere. The process of trees collecting water from the soil and transporting it to their leaves is not easy; it is a pretty mammoth task. Trees require a *lot* of moisture in order to undertake the process of photosynthesis that gives them energy to grow, but not nearly as much as they pull up.

Ninety per cent of the tree's water is released from their leaves. Under ideal circumstances, a typical mature tree has the capacity to transport approximately 10,000 gallons of water. However, only around 1,000 gallons of this water are utilized for the production of food and contribute to the tree's biomass. While this may look like another instance of the selflessness of trees, it is also an example of their long-term planning to make the wider environment and atmosphere better for them to thrive. Forest environments need clouds, moist air, rain and water recycling in order to grow.

This is a simplified explanation and by no means the only reason that trees take up so much more water than it might initially seem they need. Transpiration not only cools trees, but also aids in the movement of essential mineral nutrients and water from the roots to the upper parts of the tree. This movement occurs due to a decrease in water pressure, caused by water evaporating from the leaf stomata into the atmosphere. This continuous process ensures a steady flow of nutrients and water throughout the tree.

The next time you stroll through a forest, take notice of how the temperature changes around you. It is not just the shade cooling the environment. It is an example of how trees have evolved to control entire weather systems to support not just themselves, but the lives of other creatures they are reliant upon. On the fringes of the Amazon, temperatures can drop by up to 5°C within the forest compared to the hotter climates of neighbouring farms. Similarly, in Sumatra, Indonesia, researchers found forested areas boasting temperatures up to ten degrees cooler than adjacent palm oil plantations.

This is yet another example of how trees are so interconnected to everything on Earth. We can think of them as the guardians of all life. They pull the strings, not us. But this is not the only way trees work in harmony with other species.

Graham Alderton is not a man prone to exaggeration. As head gardener at Winkworth Arboretum, a standout in the National Trust's collection, he knows his trees. On an autumn morning, with the leaves turning rich shades of red and yellow, he walked me through the arboretum's impressive mix of over a thousand species of trees and shrubs. We stopped beneath a sprawling sweet chestnut, its twisted branches stretching skyward.

'That tree alone,' Graham said, gesturing upwards, 'probably supports thousands of species. Aphids, beetles, bats, woodpeckers, squirrels – all sorts rely on it. Even foxes and badgers feed on the acorns or dig around the roots. Then there's the fungi, worms and moss working below ground. It's not just a tree – it's a whole community.'

A single mature oak, I later learned, can support over 2,300 species – 326 of which rely on it entirely for their survival. For much of the natural world, trees like these are more than just a backdrop – they're entire worlds.

If that is what a single tree provides, think about what an entire forest can do. These buzzing kingdoms of life support countless species. The most biodiverse places of all are the tropical rainforests. When intense rainfall and warm temperature combine, it creates the perfect environment for symbiotic life to thrive. A rainforest spanning one hectare (equivalent to 2.5 acres) can host a staggering 480 tree species, a quantity 20 times greater than that found in a deciduous forest. Within that, it provides habitat for an astonishing 42,000 insect species.

Mega forests (the largest forested and most biodiverse regions on the planet) contain almost all the birds of paradise, anacondas, chimpanzees, bonobos and gorillas. Most of the planet's bugs, trees, mushrooms and freshwater supplies are in the big woods, as are hallucinogens, analgesics, tumour shrinkers, stomach settlers, anaesthetics, vision enhancers, sedatives, stimulants, and more. Life is in full riot in the intact forests. They are the planet's wildest, most biologically diverse lands.[6]

Through these symbiotic relationships, trees influenced the evolution of millions of species. As trees rose to dominate landscapes, they created a new frontier in the form of forest canopies – an aerial realm previously unexplored. For many creatures, this presented an opportunity to exploit a three-dimensional environment. With the advent of flowering trees, a fascinating mutual adaptation unfolded between trees and their pollinators. Many insects can perceive colours of the spectrum that are invisible to us. Ultraviolet photographs show that there are many more markings on petals than we are able to see.

Insects, guided by sight and the lure of scent, act as the sexual intermediaries between trees. It occurs in the most remarkable of ways. While many bugs are attracted to the sweet smell of lavender and honeysuckle, flies prefer the stench of rotting flesh. Plants respond accordingly. The maggot-bearing Stapelia from southern Africa has flowers that reek of carrion and even shapes its petals to look like the decomposing skin of a dead animal. To complete the illusion, the plant generates heat to mimic the warmth of a freshly rotting corpse.

Other plants are weirder still. One orchid even produces a flower that resembles the form of a female wasp, complete with eyes, antennae and wings – it even gives off the odour of a female wasp that is ready to mate. Male wasps, thus deceived, attempt to have sex with the flower and in doing so fulfil the orchid's will by pollinating the plant.[7]

The union between flowering trees and their pollinators stands as one of the most iconic examples of mutualism in nature. Trees offer a dazzling array of rewards – nectar, pollen or even specialized structures – to entice pollinators to facilitate their reproduction. One such remarkable partnership is the relationship between fig trees and fig wasps. They share a bond forged over millions of years.

Fig trees bear unique, enclosed inflorescences called figs, which house tiny, specialized flowers. These figs are accessible only to female fig wasps equipped with specialized ovipositors. The intricate choreography of this partnership begins when a female fig wasp enters a fig to lay her eggs. In the process, she fertilizes the tree's flowers with pollen she carries from another fig. As the wasp lays her eggs, she also unwittingly pollinates the tree. When the wasp larvae hatch, they feed on some of the fig's developing seeds, ensuring that some seeds remain to mature.

Once mature, female wasps emerge from the fig and collect pollen to repeat the cycle in another fig. Male wasps, on the other hand, remain within the fig, never leaving it but producing offspring with the females. This unique partnership ensures the reproduction of both the fig tree and the fig wasp, with each partner depending on the other for their survival and propagation. While trees may appear to be individuals, when we observe nature, it is clear that they are anything but. The fundamentals of biology and evolution are interconnection and symbiosis.

In 1973, Japanese researchers launched a unique forestry experiment near Nichinan City, Miyazaki Prefecture, aiming to observe how tree density impacts growth. By arranging Sugi (Japanese cedar) trees in concentric circles with varying spacing, they created a living study of cooperation within the forest. Now visible from above, these tree rings offer more than a fascinating geometric design – they reveal how trees respond to each other and their environment.

What emerged is a story of collaboration more than competition. Trees planted close together grew shorter, while those in the outer, less-crowded rings reached significantly greater heights. But this is not simply about competition for resources like sunlight and nutrients. Trees growing tightly together are not stunted due to selfish rivalry; instead, they appear to harmonize their growth. Rather than one dominating, they share resources, ensuring each has enough to survive. This cooperative strategy prevents the inner trees from competing aggressively, instead allowing them to coexist in balance. The result is a community where no one outgrows their neighbour, fostering a more resilient and interconnected system.

This example of Sugi trees in Japan mirrors Dr Suzanne Simard's research into tree cooperations, which shows that forests function as communal systems, with trees sharing nutrients and even supporting weaker members. Instead of focusing on outcompeting each other, trees grow in harmony, like a family, to allow all members of the forest to thrive. Trees, like many animals, live in intimate groups. Go in search of the tallest, the oldest and the healthiest trees, and you'll find them among their peers, flourishing together in thriving forests.

I found myself standing in awe of General Sherman, a colossal giant in Sequoia National Park. Rising 275 feet (83 metres), with a base circumference of over 100 feet (30 metres), General Sherman is not the tallest tree, but by volume, it is the largest known living tree on Earth. But what truly makes General Sherman remarkable is not just its volume or age. It is the tree's role as a vital part of a family – a network of life deeply connected within the forest around it.

This ancient sequoia, named after the famous American Civil War soldier, is estimated to be around 2,500 years old and exists within one of the richest, most biodiverse forests on the planet. General Sherman is part of an intricate forest system where trees work together, sharing nutrients, water and even wisdom through their root systems. Much like in a family, the largest trees in the forest act like a parent, and play a key role in supporting the growth and health of younger, smaller trees, ensuring the entire ecosystem thrives together. They are the trees that have the most connections to smaller trees through their underground mycelial networks, more than any other tree in the forest.

Like Methuselah, the General Sherman tree is older than Jesus, perhaps even older than the Buddha. Standing in General Sherman's presence, it is hard not to imagine this colossal tree being the matriarch of the surrounding forest, responsible for the care of those in its community.

Simard speaks about 'mother trees' of the forest. While not necessarily the genetic mother of all the trees in their surrounding forest, these are often the largest and the oldest trees and act as a core for vast below-ground mycorrhizal networks – like General Sherman. Their age, strength and wisdom benefit the vast number of smaller trees to which they are linked within the forest. A mother tree sustains seedlings by helping them to establish a symbiotic relationship with fungi and providing them with essential nutrients for their growth.

'Elders that survived climate changes in the past ought to be kept around,' writes Dr Simard, 'because they can spread their seed into the disturbed areas and pass their genes and energy and resilience into the future. When mother trees – the majestic hubs at the centre of forest communication, protection and sentience – die, they pass their wisdom to their kin, generation after generation, sharing the knowledge of what helps and what harms, who is friend or foe, and how to adapt and survive in an ever-changing landscape. It is what all parents do.'

Next time you are in a forest, look up. You might notice something called 'crown shyness', where some trees of the same species – and sometimes others – will avoid contact with another's branches, creating a beautiful patchwork of leaves and light. The reasoning is hotly debated by scientists. Some cite an evolutionary defence mechanism to prevent branches rubbing and thus stunting growth; others suggest it inhibits malevolent insects spreading their larvae, while others still reckon it is merely a response to shade.

Either way, trees know better than to fight with one another, and act in a way that could in any definition be described as consideration for one another's need for light. Trees can indeed make friends.

Trees have evolved to work actively together in order to survive: those in a forest even decide when to reproduce together. They prefer to do it at the same time, so that the genes of a large number of individual trees can be mixed. Trees, like animals, have survival skills, and have to weigh up their options to determine the best course of action.

Beech and oak trees provide animals like deer with a tasty meal through their beechnuts and acorns, and forests can be totally picked clean of them in the autumn, leaving no seedlings to sprout. Trees protect themselves and have their revenge at the same time. By mixing up their patterns and refusing to bloom every year, animals can no longer count on them and must endure a winter with little food. Many will starve. The next generation of trees therefore has a higher chance of succeeding, and with fewer pests left to eat their seeds.

Trees keep each other in check, too. Diversity strengthens the ecosystem of the forest. The more species there are within, the less chance there is that one species begins to dominate and overpower the others, depriving them of crucial nutrition. Even the trunk that remains after the tree is struck by lightning, or snapped off in a storm, is useful to its brethren – it manages water for the living trees through its presence.[8]

We need to start thinking about forests as an entire ecosystem more than a collection of individuals. As a forester, Peter Wohlleben had been taught that trees were competitors engaged in an almost Darwinian struggle for survival. Later on in his life, Wohlleben came across a tree stump half a millennium old, without a single leaf – cut down, ostensibly dead – and yet *alive*. The trees surrounding it had given it life support, diverting sugar solution through their roots to this seemingly forgotten stump. Instead of being competitors, it was as if these trees were a family. 'Trees,' he came to believe, 'are very interested in keeping every member of this community alive.'

In essence, trees embody a kind of ecological altruism – a benevolent giving without immediate expectation of receiving. They enhance the resilience and stability of forest ecosystems, epitomizing the intricate interplay that characterizes the natural world. Through their selflessness and acts of collaboration, trees have created a veritable garden of Eden for all creatures to not only survive but thrive. In doing so, this paradise then gave birth to a new kind of fauna that would change everything . . . us.

Chapter 4

Our First Home

———

One touch of nature makes the whole world kin.

– William Shakespeare, *Troilus and Cressida*

About 55.5 million years ago, there were palm trees in Antarctica. During the Paleogene Period, the planetary temperature rose by an extreme 6–8 degrees in a short spasm that had profound effects. The massive amounts of carbon dioxide and methane released into the atmosphere served to transform utterly the ecosystems of the world. Warm, tropical conditions spread to high latitudes, allowing species such as snakes and crocodiles to thrive near the poles. These global climate shifts prompted many animals to evolve rapidly in response to the changing environment.

Almost the entire Earth was covered in trees, and in the evolutionary blink of an eye – a matter of a few thousand years – primates appeared, as if out of nowhere. Our early ancestors made the move away from a ground-based, scent-dominated, often nocturnal existence to a life in the trees, due to the development of grasping hands, long arms, stereoscopic vision and increased brain size.[1] It took them a further 52 million years to evolve anywhere near to what we might consider hominid, but this early adaptation to arboreal life laid the foundation for what would eventually become modern humans.

For that, we can thank the trees.

There has been much debate about our origins in both theology and science. Charles Darwin was one of the most notable of the latter theorists. In 1871, Darwin wrote in *The Descent of Man* that 'man is descended from a hairy, tailed quadruped, probably arboreal in its habits, and an inhabitant of the Old World.'

Our tree-dwelling mammalian ancestors developed prehensile hands and opposable thumbs, which allowed them to grasp branches and navigate the complex canopy more efficiently. Over time, this adaptation led to the diversification of primates into various species, including lemurs, monkeys, apes, and eventually us. Look at your hands and you'll have a constant reminder that we evolved and thrived amid the branches. We will always be creatures of the forest.[2]

There is now no doubt that our evolutionary roots are arboreal, but there remains some debate about how and when we decided to leave the canopy during the last 5 million years. For most of the 20th century, the prevailing belief was that the development of our species as we know it did not begin until after we left the forests and walked out onto the grasslands. Since Victorian times, the savannah has been the crucial backdrop for the grand story of human evolution.

As Dr Patrick Roberts, an archaeologist and anthropologist, says so eloquently, 'All the while, tropical forests remained as out-dated backwaters, unattractive for increasingly upright, hunting hominins and better associated with the non-human great apes.'[3] But it has now become apparent that things were not so black and white. Having evolved for so long in the forests, it is misplaced to assume that we were so desperate to leave them behind at the first opportunity. New research suggests that perhaps humans learned to walk upright not to escape the forest, but instead to be able to travel both in and between forests more efficiently.

Gabriele Macho, a paleoanthropologist at the Catalan Institute of Palaeontology, studied the wrist bones of two of our early relatives, *Australopithecus anamensis* and *Australopithecus afarensis*. They used scans to compare these bones to those of modern primates like orangutans, gorillas, chimpanzees and humans, and found that *Australopithecus anamensis* – which is 600,000 years older than the *afarensis* – had wrist bones like those of animals that spent a lot of time in trees, while *Australopithecus afarensis* had wrist bones more similar to animals that walk on the ground, like us.[4]

Macho explained that originally hominids learned to walk upright as a means of moving quickly between trees. We can witness this movement today when we catch a glimpse of the orangutans, who often walk on two legs among the trees while grasping a branch above. If you have ever been to an adventure park and walked along the ropes, you will know that it is a motion familiar to humans.

Standing upright on branches gives the evolutionary advantage of being able to pick high-quality fruit safely in the trees, instead of scavenging for fallen fruit on the forest floor, vulnerable to predators. But soon this evolutionary trait became useful during periods of climate change and drying, as the grasslands expanded: 'With the trees being farther apart, it became energetically advantageous for hominids to cross the gaps bipedally.'

On two legs, and with the intelligence and dexterity to develop tools, our early human ancestors left Africa and began to spread out across other continents. When this happened is not exactly clear. But, the more clues we get, the more complicated the story of our evolution becomes. The emerging picture is that *Homo erectus* was the first of the human ancestors to leave Africa around 2 million years ago. In the last one million years, humans could be found all over Africa and Eurasia. Several subsequent species emerged, only to become eventually extinct. In northern Eurasia (*Homo antecessor, Homo heidelbergensis, Homo neanderthalensis* – the Neanderthals and the Denisovans), Southern Asia (*Homo floresiensis* – the 'hobbits') and us, *Homo sapiens*, in Africa.

Later species of early humans continued the trend towards increased bipedalism, gradually coming out of the trees. The famous Turkana Boy, one of the most complete *Homo erectus* specimens, dated to 1.4 million years old, shows an anatomy not too dissimilar from you and me.

Until recently, our species was considered to have emerged within the last 200,000 years in East Africa, but new hypotheses suggest a 'pan African' origin, where *Homo sapiens* emerged across diverse environments in Africa.[5]

Likewise, in recent years, researchers have also challenged the theory that early humans who had left the jungles avoided going back into them until some 12,000 years ago. The belief was that this kind of terrain would have posed too much risk compared to reward for our ancestors. And yet, the earliest known burial in Africa (78,000 years old) was discovered at Panga ya Saidi in the heart of the Kenyan coastal forest. The burial of a three-year-old child, wrapped in leaves or animal skins, shows that early humans did indeed live – and die – in tropical forests long after they were supposed to have abandoned them.

Human activity continued at Panga over the next 80 millennia, where the forest offered a stable home for our ancestors. Animal remains, seeds and mollusc shells showed that these people relied on a diverse range of resources from both the forests and the coast. Today, Panga is a tranquil connection to this past world and it remains an important place of symbolism and religious activities for locals.

Dr Rahab Kinyanjui, from the Kenyan National Museum, said, 'Tropical forests hold sacred significance in African society, revered as self-sustaining and generous ecosystems. They are regarded as Earth's detoxifying gift, often called the planet's lungs. Along the Kenyan coast, Kaya forests embody a living legacy of people's history, culture and religion, serving as burial grounds and sacred sites for rituals and prayers . . . This place is an ideal example that has preserved human culture and beliefs for thousands of years.'

Similarly, archaeological work in the rainforests of Senegal has led to the discovery of stone tools on the Gambia river, which show that people lived there at least 24,000 years ago. It seems that the forests were not abandoned at all.

Homo sapiens left Africa in successive waves. Most of the expeditions seem to have failed, but around 60,000 years ago one such dispersal succeeded. It was the descendants of this migration that went on to settle across Europe, Asia, and island-hopped to Australia – all pretty quickly. For all the range of new environments that these ancestors encountered, forests nevertheless remained an important home. By analysing carbon and oxygen isotopes in the teeth of human remains found in Sri Lanka, the oldest of which was dated to 20,000 years old, researchers discovered that the diet of these people could only have been sourced from within the rainforest.

When even older remains of shellfish were dated at almost 40,000 years old, it suggested that humans had been collecting food from the coast and rivers and taking it *back* into the jungle with them.

Dr Patrick Roberts writes that 'it was really the first evidence that humans not only could live their lives in rainforests, but were choosing to do so since their first arrival in Sri Lanka.'[6] Rather than struggling in the forest for basic sustenance, humans were becoming masters of a range of different environments.

Scientists, he explains, once believed that humans moved into much of South Asia only when the climate changed enough that the forests began to wane. One popular narrative argues that humans consistently remained close to the coast, travelling from island to island as they spread across the world.

This, Dr Roberts believes, is simply not true: 'When humans get to places like Sri Lanka, Indonesia and New Guinea, they move straight into their tropical forest environments and use them in different ways. *Hang on! We're finding microliths, human remains, and bone tools – all in rainforest habitats.* Sri Lanka is slap-bang in the middle of the Indian Ocean rim, so why would all of the earliest sites – dating back about 45,000 years – be deep in the rainforest, if human expansion depended on coastal resources?'

As our ancestors moved around the world and into new environments, their symbiotic relationship with trees continued. Wood was used for tools, such as spears, bows and arrows, and fuel for fire. The fruit of trees provided food, and their boughs gave shelter. Our ancestors had an innate grasp of life cycles, and our part within them.

Discoveries in one Sri Lankan cave revealed evidence that the early humans who inhabited the area knew that they should only hunt adult animals, so as not to diminish the ability of an animal population to replenish. In other words, our ancient ancestors understood both the importance of sustainability and how to practise it – something that our more 'developed' brains seemingly struggle with in our era of man-made extinctions.

When Charles Darwin visited the Galapagos Islands in 1835, he noted with interest that despite being hundreds of miles away from the mainland, the species that inhabited the islands bore remarkable similarities to those he had encountered before. For his theory of evolution to be correct, these creatures must be the descendants of colonizers. But the question remained – how did they get here?

So vexed was Darwin by this problem that upon his return to England he staged a series of experiments, whereby he immersed various seeds in sea water for weeks at a time and discovered that they still germinated even after considerable periods. This led him to the conclusion that plant life was indeed carried by the waves, but that did not account for the animals. How did they get there? It was not until the German meteorologist Alfred Wegener came up with the idea that 'the continents must have shifted', that we came to understand how geology impacted the way the world worked and its impact on the migration of species.

Wegener theorized that at one time 'all of the present day continents had formed one giant supercontinent – Pangea.' Now we understand continental drift, and the fossil record proves that 200 million years ago creatures did indeed once live together, and only became separated after millions of years adrift, resulting in the diversification of all animal and plant life.

The difference now, of course, is that while evolution used to take a long time, we are suddenly faced with a 'New Pangea', with hitherto distant species being suddenly brought together as humanity forges closer and closer ties. As early as 45,000 years ago, humans were transporting plants, vegetables, seeds and even animals around the world. Yams and sago palms were being transported by canoe across Oceania. Around 20,000 years ago the wayfarers were even ferrying live furry bandicoots and cuscus possums around Indonesia and Australia to ensure an easy protein snack on their long journeys.[7] For a very long time humans have been altering their landscape, introducing new species and cultivating nature to suit their needs.

Since the end of the last ice age 11,000 years ago, humans have cohabited with trees, moving into the habitats pioneered by the advancing treeline and then adapting, managing and stewarding them to create a stable global ecosystem.[8]

As Ben Rawlence describes in his book *The Treeline*: 'The ice has come and gone many times. And each time nature has begun again, slowly recolonizing the land scoured of soil. First comes lichen, then moss, then grasses, shrubs and the pioneer trees . . . left to its own devices, nature's equilibrium in most habitats on Earth unless limited by cold or drought tends towards the eventual production of forests.'[9]

The Last Glacial Maximum occurred around 19,000 years ago, with ice covering about 25 per cent of Earth's land area. There is some debate about where trees survived during this period. While the traditional view held that European tree species survived in refuges like the Balkans, a recent study suggests that spruce, pine and birch may have also survived in parts of Central and even Northern Europe. Evidence from ancient pollen and DNA studies indicates that trees were present in Scandinavia as far back as 20,000 years ago – some 10,000 years before the ice age ended.

The last retreat of the ice age glaciers marked a new era in Earth's history: the Holocene epoch. It was a period of massive growth as temperatures and sea-levels rose, and tree populations rebounded, reclaiming territories that had been buried beneath ice for tens of thousands of years. Trees began to migrate across continents, as if they were birds. As they had headed south to escape the ice, they then retreated back northwards when it passed. Soon, areas that had been veritable wastelands of snow were now teeming with plant and animal life.

Different tree species colonized these new areas at different rates. Deciduous oaks made a rapid advance, reaching southern Norway some 7,000 years ago. About 5,000 years ago, European trees reached their furthest extension northwards. Two thousand years later, beech trees arrived in Britain. One reason for this different rate of colonization was seed dispersal. Some trees, like birch and Scots pine, disperse their seeds in the wind, and can thus expand their reach much faster than the species that drop their seeds in close proximity.

Human hands began to play a part, too, assisting the spread of trees by carrying seeds of the trees that were most beneficial with them as they traversed the continent. Humans were therefore instrumental in reintroducing many types of trees to the northern regions, and in doing so, helped pay back the debt we owed them for our earlier survival.

The story of post-glacial reforestation is not limited to the highest latitudes; it resonates globally. In Europe, the return of deciduous trees like oaks and beeches transformed landscapes and enriched ecosystems. In Asia, the recovery of forests saw the resurgence of species such as the Siberian pine and the Mongolian oak, fostering biodiversity.

In South America, the Andes mountains witnessed a similar resurgence of forests, influencing the distribution of species such as bears and llamas. Meanwhile, in Africa, savannahs gave way to expanding forests, impacting the habitats of elephants, gorillas, and countless other species. The 'African Humid Period' saw an increase in monsoonal rainfall in eastern Africa, leading to the creation of large shallow lakes, swamps and mangroves.

One of the most remarkable transformations took place between 10,000 and 6,000 years ago: the greater Saharo-Arabian desert turned green. Petroglyphs – primitive rock carvings created by ancient humans – show that these massive deserts were once home to elephants, oryx and other species, providing a window into the past and the types of species that human societies would have been familiar with. Trekking along the banks of the Nile in Sudan, I encountered many large boulders with carvings of giraffes, lions and elephants in the middle of the desert.

Rainfall from the south created rich hunting grounds for the hunter-gatherers that began to call these areas their home. Geological remnants of these lakes can still be seen today in the Fezzan basin in Libya, alongside the Mundafan and Jubbah basins in Saudi Arabia. A modelling study has shown a range of tropical and iconic Sahelian (Acacia, Commiphora and Balanites) plant types which moved northwards with the increased rainfall into the western and central Sahara.[10] The expansion of trees and grasslands highlights the various migrations and movements of plant communities in response to global climate changes and nature's resilience and adaptability.

This profound ecological recovery offers insights into the ongoing challenges of climate change and habitat loss. The enduring legacy of the ice age reminds us that, even in the face of immense challenges, nature possesses the power to rebound, rejuvenate, and thrive.

A message of hope and inspiration for our stewardship of the planet, if only we safeguard our trees.

'I believe that we have rainforests in our blood,' wrote Merlin Hanbury-Tenison in *Our Oaken Bones*. 'After at least 900,000 years in the UK, our ancestors and precursor hominid species have known these [forest] habitats and lived among them for longer than we can ever hope to fathom.'

Perhaps many of the environmental questions for today can be answered by looking back to our spiritual home – the forest. As humans developed, did our ancestral connection to trees truly remain a spiritual one, and if so, what led us to forsake this vital relationship?

Act Two

Paradise Lost

Chapter 5

The Roots of Civilization

———

*Trees have always been the most penetrating preachers . . . Nothing
is holier, nothing is more exemplary than a beautiful, strong tree.*
– Hermann Hesse, *Bäume: Betrachtungen und Gedichte*

In the autumn of 2017, I went looking for the Garden of Eden.
Inconveniently located in the south of Iraq, my mission was hampered
somewhat by the fact that I was undertaking this during the latter
stages of the war against ISIS, and therefore fraught with the perils
of a brutal conflict. Having passed through the destroyed city of
Mosul, I hitched a lift with Iranian-backed mercenaries on the back
of a tank across the front lines and got shot at by Islamic
fundamentalists. In Tikrit I visited the former hideout of Saddam
Hussein – now an insurgents' bomb-making lair – before seeing the
remains of the biblical Babylon.

Having been questioned by armed militia units several times on
the road south and stopped at endless roadblocks, I was relieved to
finally make it to the small city of Al Qurna at the confluence of the
Euphrates and Tigris rivers. It was here, in this unlikely spot, on a

hot and dusty afternoon that I managed to locate the biblical paradise.

Al Qurna has been noted in travellers' accounts for centuries as the place where two of the four 'rivers of paradise' that flowed out of the Garden of Eden converge, as named in Genesis 2:10–14. Elements from the early stories of Genesis have been traced to the cuneiform tablets written by the Sumerians and Babylonians, discovered in the nearby cities of Eridu and Ur.

On the banks of the river, a wall surrounded a dusty park, which I followed to the entrance gate only to be very disappointed to discover a sign hung up outside that declared: 'The Tree of Knowledge closed for holidays.'

Undeterred, and spurred on by some local children who were playing football in the street and pointing at a hole in the fence, I snuck in through the gap and found myself in a small park filled with palm trees and concrete picnic tables. Broken shards of glass littered the floor and the whole place looked dismal. In the middle I came face to face with arguably the most revered tree in the history of human civilization.

A dead and withered stump had been secured to the concrete base. This was what I had come to see.

It was decidedly underwhelming. Today, all that remains of the tree is a pitiable stump, ostensibly a few centuries old at most, yet potentially part of a lineage in which each dead tree was replaced anew. In the latter half of the 20th century, a modest garden featuring indigenous plants was erected around this tree, intended as a reconciliatory gesture following some British soldiers carelessly climbing and breaking the tree not long after the First World War.

Early in Saddam Hussein's tenure, the area was enclosed into what is now a dilapidated concrete plaza. Under a baking sun, this ruin poignantly epitomizes the lost ideal of paradise. A dead stump in a run-down concrete shell, all that remains of the promise of Eden.

As humans settled into more permanent communities and mastered tools, language and early societies began to emerge. With these advancements came the rise of storytelling and religion. Though many had moved beyond life in the forests, humans still held deep reverence for trees, which continued to provide shelter, food, and powerful symbols in early myths and religious narratives.

In a primordial Earthly paradise there was a garden that God gave to mankind, the crown jewel of His creation. As the story goes, there was a tree in this garden, the fruit of which could grant the one who consumed it the knowledge of good and evil. All of the fruit in the Garden of Eden was offered by God to the first man and the first woman, his children, Adam and Eve. All save the fruit of this one tree. To taste that fruit meant to awaken to an understanding of the world that we were unready to bear, and to forever be denied our place in paradise.

But we could not resist, could we? We longed to taste the forbidden fruit, to open our minds to the knowledge of the divine.

The Greek titan Prometheus stole fire from the gods and was punished for his insolence by being chained to a rock to have his liver eaten by an eagle for eternity. Eve stole something altogether more sacred to them: consciousness, and the punishment was no less severe:

> *Cursed is the ground because of you;*
> *through painful toil you will eat food from it*
> *all the days of your life.*

It will produce thorns and thistles for you,
*   and you will eat the plants of the field.*
By the sweat of your brow
*   you will eat your food*
until you return to the ground,
*   since from it you were taken;*
for dust you are
*   and to dust you will return.*

So said the Lord to Adam, as punishment for eating the fruit from the Tree of Knowledge.

He gave Eve an equal remonstration, stipulating that because she listened to the snake and tempted her husband, she would suffer pain in childbirth and forever be subservient to man. The snake got let off lightly by being ordered to slither on its belly for eternity and being fair game for having his head bashed in.

Eve's decision transcends mere disobedience. The fruit signifies not just forbidden knowledge, but also speaks to the eternal human aspiration to reach beyond the familiar, into the unknown, to seize what seems just beyond our grasp. This portrayal has profoundly influenced the cultural narratives of Christendom, positioning the apple as a potent symbol of desire, not solely for power but for the enticement of the forbidden.

Two hundred kilometres west of the Tree of Knowledge in Iraq lies the ancient Mesopotamian city of Eridu. Nothing much remains now apart from a small hillock that was used for target practice by the American troops who invaded in 2003. When I visited, it was unoccupied except by a couple of bored-looking Iraqi soldiers, who told me not to steal anything. I promised I would not.

Walking across the crumbling remains, it was hard to imagine that this was once a thriving metropolis, perhaps the oldest city in the world. It dates back to before the Great Sumerian Flood, to pre-biblical times, and archaeological evidence suggests its origins go back at least 7,500 years – a full 3,000 years older than the pyramids of Giza.

Everywhere I walked, the crumbling remains of ancient pottery and bricks littered the floor, and then I realized that these were not any old bricks. On closer inspection these were cuneiform tablets, the sort to be found in the British Museum recording the Epic of Gilgamesh and the Great Flood. There were thousands of them, broken, lying all around with their tiny inscriptions.

The cuneiform script is the oldest written language in the world. And it came at a time of great change. As neolithic hunter-gatherers took up farming and began to settle along the banks of the great rivers, they needed a way of recording their transactions. Money was invented, and it should come as no surprise that the first written words were simple accountancy etchings – ledgers to record the transactions of grain.

Here, on the dusty plains of southern Iraq, I had come looking for the Garden of Eden, and yet found only the remnants of an accountant's bill.

Adam and Eve's eviction from the Garden of Eden symbolizes humankind's loss of innocence and its disconnect from nature, which would come to be understood as a tragedy. Leaving paradise – the forest – and being forced into agriculture to survive marked an enormous shift in the evolution of humanity. After millions of years of evolution (condensed into a single sentence in the book of Genesis), where humans had lived almost exclusively in forests as hunter-gatherers, the period after the last ice age heralded the dawn of a new era.

Stories rooted in trees are woven throughout ancient history, offering insights into humanity's evolving relationship with nature. As humans transitioned from forest dwellers to agricultural societies, by looking back at these myths and legends we can get a glimpse into the fading connection between mankind and the forests – a shift that would shape their future.

The pervasive myth that the forbidden fruit of the Garden of Eden was an apple chiefly originates not from biblical text, but from a blend of linguistic nuances and cultural reinterpretations. At the heart of this misconception is the Latin word *malum* which, somewhat confusingly, serves as a homonym for both 'apple' and 'evil'. This fateful linguistic confluence has profoundly influenced the artistic and literary portrayal of the forbidden fruit in Western Christian tradition.

This iconic depiction has been carried forward through the ages, enduring in cultural memory despite the absence of any specific reference to an apple in the original Hebrew manuscripts. A curious case of fate has cemented the apple as a potent emblem of temptation and desire within the collective Christian imagination, and in doing so has brought it much closer in symbolism to other ancient stories about apples.

Apples appear often in the old myths and fairy tales, and almost always in relation to desire, if not of the flesh, then of the heart. In the classic fairy tale 'Snow White', the Wicked Queen disguised as an old hag – another potent universal archetype in the world's folkloric traditions – offers a poisoned apple to Snow White. Herself the picture of Edenic innocence and naivety, Snow White bites the apple, falling into a deathly sleep. In the story, the apple is imbued with symbolic meanings of desire and loss of innocence.

The Hellenic Aphrodite was the goddess of love, and she is often symbolized holding an apple. How she acquired that apple is a myth to rival that of Eden. As the story goes, there was a wedding feast among the gods of Mount Olympus, only one of the gods was not invited. Her name was Eris, the goddess of discord, and the way she handled this situation was a credit to her title. Denied an invitation to the party, she decided to make it miserable for everyone else. Sneaking into the party, she tossed a golden apple into the crowd. Inscribed upon it were the words 'to the fairest'.

Naturally, several goddesses assumed the apple was for them. Hera, Athena and Aphrodite all quarrelled over the apple, and Zeus finally had to step in. Zeus appointed Paris, the Prince of Troy, to determine who was the fairest goddess of all. This idea of a human mediator for a divine dispute is a common one in Greek mythology, and demonstrates that, for the Greeks, there was less of a clear delineation between the human and the divine than there were for later religious traditions.

Paris, far from being an impartial observer, allowed himself to be plied with bribes from the three goddesses. Athena, goddess of war and wisdom, offered him unparalleled skill in battle. She promised to turn him into the greatest warrior in all of Asia Minor. Hera, for her part, simply offered him all of Asia Minor. But it was the goddess Aphrodite's bribe that ensnared young Paris. She offered him the hand of the most beautiful woman in the world, and he could not resist. He bequeathed the golden apple to Aphrodite and declared her the fairest of them all.

Not long after, Paris was secretly married to the most beautiful woman in the world. The woman whose face 'launched a thousand ships'. Her name was Helen, and the ships that embarked because of her beautiful face formed the Greek invasion fleet that set sail to lay siege to Troy. The goddess of discord and the goddess of desire had accidentally conspired to bring about the Trojan War. Once again, the apple of desire was to blame.

Early stories featuring trees and fruit symbolize humanity's once deep connection to nature. They warn of the dangers of impulsive decisions driven by desire, highlighting how this disconnection from nature leads to imbalance. These ancient myths remind us that we were once in harmony with the natural world, a connection many have forgotten but urgently need to restore.

The Oxford English Dictionary, that crowning achievement of British peevishness, defines nature as 'the phenomena of the physical world collectively, including plants, animals, the landscape, and other features and products of the earth, as opposed to humans or human creations.'

What immediately leaps from this definition is the explicit assumption that mankind is something inherently separate from nature, and vice versa. It reinforces the arrogant concept that we exist apart from nature and have dominion over it.

But where did we develop this urge to quell the forests, to subjugate nature to our will and utilize her gifts strictly for our own ends? Here, we must return to the Garden of Eden. There have been many disastrous outcomes resulting from mistranslations and misinterpretations of Holy Scripture – from justifications for the transatlantic slave trade to arguments for the inherent inferiority of women and the radical notion of inherited, original sin – but perhaps among the most disastrous is a mistranslation of Genesis 1:26–31:

> *And God said, Let us make man in our image, after our likeness:*
> *and let them have dominion over the fish of the sea, and over the*
> *fowl of the air, and over the cattle, and over all the earth, and over*
> *every creeping thing that creepeth upon the earth.*

The operative word here, of course, is 'dominion'. In its contemporary usage, the word is invariably associated with the idea of supremacy and control – to 'dominate' is to lord over and take pride in the power one holds over another thing. 'Dominion' conjures an image of violent subjugation and submission, the bending of weaker things to a stronger will. The Bible also tells us that humankind's responsibility towards God's creation is 'to till it and keep it' (Genesis 2:15). But how that 'keeping' is done, has long been a matter of debate.

For some, these passages represent a charge to shepherd the natural world with care and intention. For others, they represent a licence to exploit nature and do with it as we please. This speaks to the heart of an historical conflict between Western notions of dominion over nature and indigenous understandings of cooperation with it. Unfortunately, the former disposition seems to have prevailed historically.

The problem is, this inherited belief about the implications of Genesis 1:26–31 stems from a bit of mistranslation. The word 'dominion' here is derived from the Hebrew word *yirdu*, and many Hebrew scholars concur that it is a poor translation at that. The word does not have a direct English equivalent, and it conveys the idea of a sort of suspension, wherein humanity's role is in constant tension between rising to the level of the angels and falling below that of animal life. As such, *yirdu* conveys something closer to the idea of 'coming alongside' of nature, rather than ruthlessly subjugating it to human designs. A more accurate definition of the term, then, might be *stewardship*.[1]

The Hebrew Bible is filled with other, more positive messages about environmental stewardship, and some see Judaism as one of the world's earliest environmental religions. Rabbi Norman Lamm of Yeshiva University reflects in his book *Faith and Doubt* that the religion of Judaism 'possesses the values on which an ecological morality may be grounded.'[2] One of the basic commandments of Jewish law is *bal tashchit* – do not destroy. It is derived from the twentieth chapter of the Book of Deuteronomy, and commands the Israelites not to cut down fruit trees to assist in a siege.

Even within Christianity, which largely bears the blame for this overemphasis on domination of the natural world, there is a theological tension between stewardship and dominion. In 2015, when Pope Francis released his controversial encyclical *Laudato Si*, it contained a stirring condemnation of environmental degradation, ecological devastation, and the reckless use of fossil fuels; all grounded in a deeply theological understanding of mankind's role as the stewards of nature, not the conquerors of it.

Religious and cultural texts from different periods carry subtle warnings about this increasing separation, and their messages still echo today. As we look back, we see how this relationship has shaped civilizations and, in turn, been shaped by them.

The dominant Judeo-Christian myth of the Garden of Eden and the tree that dooms those who eat its fruit has, in many ways, defined our unsteady relationship with the natural world. But looking outside of Eden, what do the world's spiritual traditions have to say about trees? Are they marked by the same broken relationship?

For untold millennia, trees have served as a guide for humans, revealing to us the hidden truths concealed within nature's rhythms. Six thousand years ago, frankincense was being traded across Arabia. It offers no sustenance except its holy odour. From the sacred olive groves of ancient Athens to the solitary oaks that stood as watchful sentinels over the druid rites of Celtic Europe, trees have long been revered not merely as sources of beauty, food and shade, but as symbols of wisdom and a connection to the divine.

The Garden of Eden appears in all the Abrahamic religions, but in Islamic tradition, the main tree is the Tree of Eternal Life, known as *šajara al-ḫuld*. Like the Tree of Knowledge, this tree comes with a warning, but instead of revealing good and evil, it holds the secret to eternal life. This reflects another deep but dangerous human desire – the quest for immortality.

It was beneath the Bodhi tree that Siddhartha Gautama awakened to the narrow way to escape *samsara* – the ceaseless cycle of birth, life, death and rebirth – and, in doing so, became the Buddha. The Hebrew YHWH spoke to the prophet Moses through a great many mundane vessels, but the first was a tree: the burning bush. It was a tree that bore aloft the crucified Christ, atop the hill at Calvary, and when Christian missionaries reached the shores of pagan England, they found it was trees that vied with their Lord for spiritual dominance of the Emerald Isle.

It is clear that in these early stories, trees are more than just physical entities; they are rich symbols imbued with spiritual significance across various religious traditions. They represent profound truths about human existence, the quest for understanding, and the search for eternal life or enlightenment. But they also warn us of the dangers of disconnecting from nature and thus our understanding of ourselves and the universe.

In the Qur'an, verses are called *āyah* or signs of God. 'Have you not considered the extraordinary bounty of nature,' the book asks. The Qur'an presents nature as Allah's prime miracle filled with signs of his compassion and concern. Muslims are encouraged not to take nature for granted and to be constantly aware of the 'signs' of God. We are instructed that wherever we look in the natural world, we find a revelation of the divine. 'Wherever you turn there is his face. God is all pervading, all knowing.'[3]

Wandering the realms of religion renders insights into an idea at the heart of many great mythological traditions: that the tree bridges the realms of heaven, Earth and even the underworld.

The recurrence of the *axis mundi* (also known as the 'tree of life') in the religious imagination of the world's great traditions is an example of what the writer Joseph Campbell would call the Monomyth. The tree, whether it be the Cross of Christ, the Norse Yggdrasil, Jacob's Ladder or the Islamic Tree of Immortality, represents the World Navel, the central place where all duality is reconciled.

Trees feature heavily in Greek mythology. In Homer's *Iliad* and the *Odyssey*, heroes are likened to trees in abundance. When Hector and Patroclus are in battle, their war cries are compared to the noise of the 'winds, striving with one another in shaking a deep wood in the glades of a mountain – a wood of beech and ash.' When the defenders of Troy are being besieged, they 'stood firm like oaks of lofty crest on the mountains . . . firm fixed with roots great and long.'[4]

Only by traversing the *axis mundi* can the shamanic Hero of Joseph Campbell's Monomyth acquire knowledge of the outer world and bring it back to save the people. In *The Divine Comedy*, Dante must descend into the inferno and ascend into paradise in order to return with the insight that will change the world. Jack must climb the beanstalk and brave the giant in order to return with the goose that lays the golden egg. So must we traverse the *axis mundi* in order to attain what Campbell referred to as the 'Elixir of Life'. The Knowledge of Good and Evil. The Secrets of Immortality. Tree, ladder, pole or beanstalk. The form of the *axis mundi* is less important than the function it plays within this singular, mythic story.

The eternal symbol of the tree as connective tissue between this life and the next dissolves that old dichotomy between stewardship of the Earth and dominion over it that we find all the way back in the book of Genesis. In the concept of the *axis mundi*, we see that our role in the natural world is not to rule over it or simply use it, but to care for it as gardeners. We are invited to work with nature, understanding that it influences us just as much as we influence it.

The date palm's connection to the divine is a profound example of tree symbolism in religious traditions. While technically a grass, its cultural and spiritual significance cannot be overlooked. Each year, Christians commemorate Palm Sunday, marking Jesus' entry into Jerusalem with fronds laid at his feet. This ritual is rich with meaning, culminating in the burning of the fronds to create ashes for the following Ash Wednesday, a practice that reflects the hope of a forthcoming Kingdom of Heaven.

Emerging from desert landscapes, the significance of palms resonates deeply with the Abrahamic faiths. In such harsh environments, these trees symbolize life and the vital presence of water, acting as markers for oasis communities. Jesus' promise of *hydōr zōēs*, or Water of Life, resonates powerfully in these arid contexts. Similarly, the Qur'an recounts Mary finding solace beneath a palm tree during childbirth, evoking the lush gardens of Jannah – paradise itself. The association with palms extends back to ancient Egypt, where goddesses like Nut and Hathor embodied water and nourishment, further cementing the palm's role as a symbol of life and hope.

The prevalence of the *axis mundi* idea is also often evident in cultures that share a shamanic belief system. Shamanic cultures tend not to make the types of dualist distinctions between the world of things and the world of spirits.

The notion of trees as central pillars connecting the celestial with the terrestrial is a deeply ingrained symbol in Mesoamerican mythology. The World Tree of Maya, Aztec and Olmec mythology, often depicted with its branches supporting the sky and its roots delving into the underworld, illustrates the Mesoamerican belief in a cosmically interconnected universe, where the physical and the spiritual meet. It is no wonder then that trees are sought after to provide wisdom and guidance.

In the Bhagavad Gita, Lord Krishna proclaims, 'Among the Trees, I am the Ashvattha,' or Sacred Fig. For Buddhists, one particular fig has earned a special significance. In 589 BCE, after many years of wandering, the 35-year-old Buddha found himself at the base of a great fig tree in Bodhgaya, a small town in what is now north-east India. He started to meditate and fell into a deep trance. Forty-nine days passed whereupon he neither ate nor slept until he became enlightened. Buddha continued to sit under the fig tree for a further seven weeks in continued meditation. After that he gave his first lecture on 'the middle way', the path to realization, and then carried on his journey, teaching.

For Buddhists, this fig became the 'Bodhi tree' and acquired a holy significance. However, due to the tumultuous periods of Mughal invasions in northern India, the original Bodhi tree of enlightenment was destroyed by Islamic armies. Nevertheless, before it was completely destroyed, a cutting of the tree was spared and retained by King Ashoka. It made its way to a temple in Anuradhapura, Sri Lanka, more than 2,300 years ago, and to this day is the oldest continually cultivated tree in the world. For millennia, devotees have come to this place in the hope of discovering the divine. Heaven on Earth.

Enthusiasts of Nordic history will recognize the familiar shape of Yggdrasil, the great World Tree of Norse mythology. Yggdrasil acts both as a narrative device and, in many ways, a central character of Norse myth. The World Tree connects the nine realms, joining the human-dominated Midgard – the 'middle earth' – with Asgard, the realm of the gods; Jotunheim, the land of the giants; and Helheim, home of the dishonoured dead. The dwarves get their own world too, as do the elves. Connecting all of these are the cosmic branches of the Great Ash, the World Tree Yggdrasil.

Elves, dwarves, trees and giants. If this is all starting to sound Tolkienesque, that is because the author J R R Tolkien was heavily inspired by the Norse myth cycle when he wrote *The Hobbit* and *The Lord of the Rings*. Tolkien's view of trees was deeply reverential, seeing them as ancient, living beings complete with their own personalities and stories. Within the world of his Middle-earth, trees are not inanimate objects or mere unfeeling organic life. They speak, they breathe, they even *move*. Trees leap from his stories as resilient guardians of ancient wisdom. When they die or are displaced, or chopped down for firewood, the world groans at the loss of its ancient sentinels.

Even from within a predominantly Catholic worldview, Tolkien could not avoid a hint of tree worship. In his masterful compendium of Middle-earth's mythology, *The Silmarillion*, Tolkien plants deep within his creation myth the Two Trees of Valinor, Telperion and Laurelin, symbolizing the light of the Valar, the gods of Middle-earth. Like the Norse Yggdrasil, they serve as the *axis mundi* of Tolkien's fictional world, and their lives and fates are deeply intertwined with Middle-earth's own destiny.

Unlike Tolkien's Valar, the Norse gods, the Æsir, use Yggdrasil to traverse the nine realms. Odin, the primary god of Norse mythology, even sacrifices himself upon the World Tree to gain knowledge of the other realms. Plucking out his eye and casting it into the well of the god Mímir, he impales his body upon his spear and hangs himself from the branches of Yggdrasil for nine days and nine nights.

His words are chronicled in the epic poem *Hávamál*:

> *I ween that I hung on the windy tree,*
> *Hung there for nights full nine;*
> *With the spear I was wounded, and offered I was*
> *To Othin, myself to myself,*
> *On the tree that none may ever know*
> *What root beneath it runs.*
> — Bellows's translation, *Hávamál, v139*

A god pierced by a spear, sacrificing himself *to himself* upon a tree. Where have we encountered this before? It seems that once again, the tree story emerges alongside a strong association with divinity. In another echo of Abrahamic tradition, the dragon Níðhöggr gnaws the roots of the World Tree, hoping to destroy it. He knows that if Yggdrasil falls, the delicate order of the Norse cosmos will fall back into primordial chaos, marking the dawn of Ragnarok.

There is that warning again: if the tree dies, the world falls into chaos. It seems even the Vikings understood, albeit through the language of myth rather than that of science, that the forests guard the fate of man, and without them, we are lost.

So while many early religious stories and myths still warn us of the dangers of a disconnection to nature, many others also demonstrate the huge extent to which early civilizations worshipped and revered her. As humans evolved, they began to do so with a respect for trees and one could argue, with a slight fear of what would happen if they were to anger nature and endure her wrath. This was a turning point in history when some started to do just that.

We modern people like to think of ourselves as rational, scientific creatures, who have gotten beyond childish myth and empty ritual. But the religious instinct in man is a far older thing than modern notions of scientific objectivity. Deep within our genetic memory there lies a dormant but unvanquished impulse to set up 'images and groves in every high hill, and under every green tree'.[5] There seems to be a part of us that stems from this early human history that longs to worship the trees, or through them, to worship what Dante called 'the love that moves the sun and all the other stars'.

But trees are not always mere spiritual symbols. They often appear as essential building blocks in the foundations of society.

The Christmas tree, now a beloved tradition in Christian and non-Christian households, finds its roots in ancient pagan practices. Long before the advent of Christianity, ancient civilizations like the

Egyptians and Romans used evergreen trees, wreaths and garlands to celebrate the winter solstice, symbolizing eternal life. In Norse and Germanic traditions the Yule tree was a symbol of rebirth and the return of the sun.

In the Christian tradition, it evolved to represent the tree of life from the Garden of Eden, as well as the promise of eternal life through Jesus Christ. However, the Christmas tree, like many modern traditions, may also reflect an unintended separation from nature. Once a symbol of the cyclical renewal of life and deep connection to the earth, it is now often cut down, adorned artificially and discarded after the season. In this way, it subtly warns us of the growing disconnect between humanity and the natural world, as we take part in rituals with diminishing recognition of their original, earth-bound significance.

This ancient reverence for trees, seen in both Christian symbolism and pagan traditions, extends far beyond the Christmas tree.

The cedar gets no less than one hundred mentions in the Bible and is acknowledged in the Psalms, where it is said: 'The righteous flourish like the palm tree and grow like the cedar in Lebanon.'

Cedars occupy a revered spot in many of humanity's mythological narratives, even as far back as some of our earliest written sources. The ancient Sumerian Epic of Gilgamesh recounts the protagonist's venture into the Cedar Forest, a realm associated with the Sumerian gods, to defeat the monstrous guardian Humbaba in a quest for immortality.

While the earliest versions of this story place the divine Cedar Forest somewhere in modern Iran, later sources position it in Lebanon. It is here that the strongest historical and mythological references to cedars find their origin.

The love poetry attributed to the biblical King Solomon often references the Cedars of Lebanon – usually with barely veiled eroticism – as comparable to the physical attributes of his lover. The Hebrew Bible describes how Solomon sourced the wood for his temple from these cedars, in yet another example of the undying human association between trees and divinity. Other legends state that these trees also provided the infrastructure for the building of the pyramids at Giza.

I had the opportunity to view some of these cedars when I travelled to Lebanon on an expedition across the Holy Lands. This much-diminished pocket of heavenly snow-capped trees exists largely thanks to the conservation efforts of Queen Victoria of Prussia, who in 1876 was concerned at the lack of care given to this historic wood and paid for a high stone wall to surround the 250-acre grove.

In North America, the western red cedar (*Thuja plicata*) is profoundly significant to the indigenous peoples of the Pacific Northwest, stretching from coastal Oregon to south-east Alaska. Known in the region as the 'tree of life', this cedar is central to many facets of both daily living and spiritual practices. Western red cedar trees provide materials for an astounding array of uses. Its timber is essential in the construction of dwellings, totem poles, ceremonial objects, masks, utensils, containers, planks, musical instruments and canoes.

For thousands of years, the coastal First Nations of British Columbia have utilized both red cedars and yellow cedars, valuing them not only for their diverse material qualities but also for their spiritual essence. The harvesting of both cedars is performed with great care, in a process designed to honour the living spirit of the tree. This profound respect for trees as living beings highlights a connection that is often found between shamanic and totemic communities and their natural environments.

All around the world, trees have come to symbolize the importance of humanity's intrinsic connection with nature. In the old Grimm fairy tale, 'The Spirit in the Bottle', the mercurial life force is discovered in a sealed bottle, hidden among the roots of a great oak tree, which extend into the eternal realm. The alchemical philosopher's tree is also rooted not in the earth but in the heavens. Symbolically it conveys the idea that the treelike process of growth in which consciousness is transformed originates in the everlasting dimension.

The oak is perhaps the most venerated of the great trees. Evoking what is regal, solid and powerful, oaks often feature in myth and legend. The Roman poet Virgil claimed an oak tree gave birth to the first humans, just as Norse Gods whittled Embla, the first woman, from an oak tree. Zeus and Jupiter were associated with oaks; Athena gave Jason and his argonauts a living, speaking oak for the hull of his ship, and the ancient oak cult heralded the communications of the oracle. Oaks are wise, reliable, and homely. They are the places where heroes go to seek knowledge, and occasionally hide from their enemies.

Whether as a sly fox with a dashing green cap, the unsmiling grimace of Russell Crowe, or in the infamous green tights worn by the likes of Errol Flynn, Robin Hood is a cultural touchstone for many. Robin and his band of Merry Men have transcended their origins in English folklore and spawned an endless series of book, film and stage adaptations, some more serious than others: Mel Brooks's irreverent *Robin Hood: Men in Tights* springs to mind.

While the story changes its shape and form with every new telling, the basic ingredients of each adaptation involve an outlaw named Robin, his outlaw band (the infamous Merry Men), a wicked old Sheriff, and of course, a forest. *The* forest – the Great Greenwood. *Sherwood* Forest. Perhaps one of the most famous forests in all of history and literature.

I happened to grow up barely an hour's drive from Sherwood Forest, or what is left of it anyway. As a boy from the West Midlands, I grew up surrounded by the ever-present myth of Robin Hood, but mostly as a convenient snare for tourists, who would only rarely find themselves as far to the north of England as Nottingham. Robin, his band of men and the forest at my doorstep were an inescapable reality.

As far as we know from the historical record, there was no real Robin of Locksley. If there was one, he does not show up in the administrative records of Nottinghamshire and, considering that he is such an iconic symbol of British folklore, he makes a minimal imprint on the pages of history and legend until the 14th century, a full two centuries after his supposed life in the divided England of King John.

As an historical figure, Robin Hood, if he did exist, is perhaps less interesting than his counterpart of legend. The earliest folk tales claim he was a commoner, a yeoman, and his benevolent regard for the poor, as well as his aristocratic pedigree, are later editions. The folkloric Robin Hood is a trickster figure, capable of many guises, and identified wholly with the forest. He dresses all in green, and springs from the trees to ambush tax collectors and Sheriff's men.

This Robin Hood almost does not seem like a man at all, more a creature of the forest. Something indigenous to it, and capable of disappearing back into it in the blink of an eye. You see, the Robin

Hood of legend is essentially another manifestation of the omnipresent Green Man.

This mystical symbol emerges as something of a recurring figure in various English folkloric traditions. The Green Man can be found in many different guises; sometimes bearing antlers appearing from his mischievous head and foliage sprouting from a grinning mouth, with a face often covered in leaves.

The adjacent folklore of the Oak King and Holly King of pagan tradition shares this idea of a personified man of the wood, who governs the waxing and waning seasons and represents a wild hidden spirit of the untamed wood. Perhaps the Green Man is a homegrown pagan symbol, or a natural representation of the Christian figure of the Holy Spirit, or maybe he is an imported symbol from a pagan tradition in a distant land. Mostly likely, he represents some amalgamation of all three.

The Green Man is carved on a tomb in France, dating back to 400 CE, while leafy-haired Gods can be seen in ancient Greek and Roman mythology and art. He is found all over the world; from Lebanon and Iraq 2,000 years ago to an 8th-century Hindu shrine in Rajasthan. He appears on temple walls in Pakistan and in churches in Borneo. In more recent times, he was even depicted on the invitation for the Coronation of King Charles III and Queen Camilla in 2023.

Chivalric romances like Sir Gawain and the Green Knight, tales like Robin Hood, and the endemic 'Green Man' pubs that can be found in practically every town in England are all tendrils of a mysterious figure that emerges from the greenwood. As an icon, the Green Man is most commonly depicted alongside oak leaves. This is no accident. Throughout history, the oak tree has served as a symbol of strength and endurance.

When Julius Caesar sent his armies into Germany, they reported back that there appeared no end to the oak forests. The Emperor observed the tribes in Germania venerating groves of trees and oaks as sacred symbols. According to a 9th-century German chronicler, when St Boniface, the Anglo-Saxon Christian missionary, arrived there more than half a millennium later, he found the local population still worshipping beneath great oaks. He cut down one of them, known as Donar's Oak – or Thor's Oak – and built a church with it upon the site where it stood.

Again, we see another example of how scratching the surface of Christendom reveals the echoes of a pagan past in which tree worship was pre-eminent, both within Britain and beyond it. The Celtic word for Druid – *Duir* – means oak. The druids, the priesthood of the Celtic tribes of Britannia, were literally 'those who know the oak'.

Robin Hood's fight to protect the Greenwood from corrupt authorities reflects an older reverence for nature, rooted in ancient belief systems, at a time when forests were still seen as communal sanctuaries. But the legend also foreshadows the encroaching disconnection between humanity and the natural world, as land was increasingly enclosed, owned, and exploited – setting the stage for a future where nature would be something to be conquered, rather than revered.

When the Scottish poet Sir Walter Scott set the fictional meeting between King Richard I and Robin Hood in Sherwood Forest in his novel *Ivanhoe*, little did he imagine that he was creating perhaps the most famous, imaginative historical forest in the world.[6] One tree, known as the Major Oak, still stands today in the heart of Nottinghamshire, a living link to the folkloric association of oak trees with protection and resistance.

Another huge white oak tree called the Charter Oak, near Hartford, Connecticut, got its name from a legend about Hartford residents hiding the Royal Charter of 1662 inside it to prevent the Charter's confiscation by a new Governor-General, who sought to invalidate the extensive freedoms it granted. The image of the oak can still be found circulating bar-room pool tables and jukeboxes, emblazoned onto the face of the Connecticut state quarter dollar.

During the Middle Ages, governing assemblies in Spain's Basque Country would convene beneath great oaks. One of these, the Guernica Oak, became a symbol of Basque identity and political self-determination. It stood through the Carlist Wars in the 19th century and when Falangist troops sought to cut down this symbol of Basque liberty during Franco's dictatorship, a group of separatists from the Basque heartland of Biscay set up an armed guard to prevent them from doing so.

Stories such as these leap from the pages of history and folklore, forming an image of the oak not merely as a majestic natural marvel, but also as a symbol of protection, strength, and liberty. The repeated integration of oak trees into folk history demonstrates the connection that human beings have shared with the oak since time immemorial. There are other species too that deserve recognition.

Walk into almost any old churchyard in Great Britain and you will most likely find yourself in the presence of a gnarled and twisted ancient yew tree, giving shade to the gravestones of those buried beneath. The yew has long been praised as a symbol of longevity, possibly because they are notoriously difficult to date with any accuracy. These poisonous trees keep their leaves through the dark winter months and have an astonishing ability to regenerate. Felled and seemingly rotten old trees frequently spring back to life and stems can root in contact with the soil, earning them the moniker of the eternal, or immortal tree.

Technically they *could* live forever, sprouting new trunks from their hanging branches in a form of slow-motion, eternal gymnastics. Perhaps this is why the Celts viewed them as sacred. Christian churches were often merely the latest in a long line of buildings located next to the powerful yew. Many of these ancient trees date back thousands of years, well before the Normans brought their stonemasons and even before the Romans started building their temples. We associate the yews with churches, but they are the vestiges of a much more ancient form of worship. The yews came before the church.

One such place that is home to yew trees and represents the unique blend of pagan and Christian culture intertwined is the famed Rosslyn Chapel in Midlothian, Scotland. You might recognize it from the movie or the book *The Da Vinci Code*. It is the setting for the climax – the discovery of the Holy Grail. While that might be consigned to the realm of fiction, the interior of the chapel is no less wondrous.

Crossing the threshold of the church, you are immediately struck by the ornate carvings that adorn the ancient doorway. Bold statuettes of Christian saints guard a massive rose window, prominently displaying a cross, framed by an interlocking pattern of *fleur-de-lys*. A stone relief features an image of Christ, not on a cross, but nailed to a tree. The inscription reads: 'I am the True Vine and Ye are the Branches.'

Stepping inside, the mystery continues. Symbols of nature and leafy motifs bedeck the soaring stone arches, which twist and writhe as though they have been carved out of living rock. You feel as if you have just passed not through a church doorway, but through a portal to an extraordinary new dimension. You are, all of a sudden, cradled in a realm where medieval mysticism and spirituality intertwine with exquisite detail. Dappled light streaks through stained glass windows, illuminating the stone in dazzling colours, bringing the carvings to life. Faces, strange and grotesque, interwoven with foliage on pillars and leering down from high arches; a hundred green men hiding in plain sight – the religion of nature alongside the worship of Christ in all its collective glory.

There is another yew at Rosslyn Chapel, but this one is a stone carving. The chapel's famous 'Apprentice Pillar' symbolically represents Yggdrasil, the World Tree from Norse cosmology. In the soaring stone arches, knotted with lifelike cusps and spandrels, with their tree-like carvings, one message is clear: the trees have always been with us.

Modern Britain is largely a secular place, but to enter Rosslyn Chapel is to take a journey deep into its Christian heritage and then beyond, into its pagan past and back further still, to those early humans who were so connected to their forest homes. But sadly, while trees are often venerated in myth and legend, they are not always accorded the same respect in reality.

As human society settled, the branches of civilization began to cast a long shadow. In doing so we began to disconnect from its roots.

Chapter 6

Broken Boughs

———

You may expel nature with a pitchfork but she will always return.

– Horace, *The Epistles*

I was 18 years old when I set out on my first trip to explore the rainforests of the world. My grandfather, something of an expert on the jungle himself, had a few words of caution for me before my departure.

'Watch out for the leeches,' he said, with a wise glare.

My grandfather had been 18 years old too when he saw his first forest. I imagined his sense of excitement, trepidation and fear as he boarded a ship bound for India and set sail for the first time. Our reasons for travel could not have been more different: he was a young man being sent to war. I was looking to understand and experience a part of the planet that I had only seen in books and in television, and to immerse myself in an environment at once unfamiliar and yet enticingly homely. At least I did not need to worry about booby-traps and snipers where I was going. Just the leeches.

Of course, in the hubris of youth I paid lip service to my grandfather's advice, as he had no doubt ignored the well-intended warnings of the experienced when he set out on his own journey. It is often the case that we need to learn the lessons of our ancestors ourselves, first-hand. Over the years I would come to find out that this is true not only as individuals, but as generations and even as a species.

From Australia, I travelled to the jungles of Cambodia and Thailand, where the famous Great Hornbill with its distinctive wing-beat can be heard overhead. In the pine forests of India, I followed the tracks of bears and wolves, and somewhere in the steamy jungles of Nepal I imagined the lowly growl of a Bengal tiger.

It was an exciting adventure, every forested mile bringing some new incredible experience. If only I had paid more attention to my grandfather's advice; the bloodsuckers got me at every turn. Every hour I would open the laces of my boots in those damp foothills to be greeted by a torrent of bright red blood oozing from my socks. It was a grisly sensation that I will never forget. My grandfather always said that leeches were the most dreadful part of the war. Coming from a man who experienced hand-to-hand fighting against the Japanese army, that is a powerful statement.

As annoying as it was dealing with parasitic worms, I felt that being a banquet for the insatiable invertebrates seemed a worthwhile price to pay for the experience of returning to the environment of my ancestors. Unsatiated, it was the beginning of a lifelong fascination with an environment that was in equal parts impossibly complex and soothingly simple. Some years later when serving in the British Army I was excited to learn that I was being sent to the small country of Belize in Central America to learn to survive and soldier in the jungle.

In many ways my time in Belize was everything that I had hoped for as a child, when I used to hack through the 'jungle' of my grandfather's back garden. Local Maya trackers showed us how to make traps and hunt for bush pigs; they taught us how to make fire and shelter using only the materials we could find on the forest floor. We learned which plants were edible and those that would kill us.

And then there were the *real* lessons, many of which came at night. At sunset, the mosquitoes arrive in their droves, biting through clothes and through the holes in a hammock. Nobody fights or even moves in the jungle at night – it is too dangerous; you will end up breaking a leg falling over or drowning in a swamp. It is pointless to send out sentries, because the vegetation is too dense, and the noises of crickets and frogs drown out every sound anyway.

All one can do is try to get some sleep amid the cacophony and hope the enemy – real or imagined – is as sensible as you. Bats swoop low and the ground is alive with movement: centipedes the size of your forearm and tarantulas as big as a hand crawl inches below your swinging body. At that point you try to remember if you had the foresight to hang your boots from a branch to stop the critters making a home of them. But even the branches are no refuge from the hordes of fire ants, or the deadly fir-de-lance viper, whose bite will kill a man within a few hours if left untreated.

Then there are the diseases which, though less glamorous, are equally likely to be the cause of your demise. Malaria, of course, which my grandfather caught several times during his sojourn in the tropics and suffered with for the rest of his life; it is thought to have killed at least half the people that have ever lived on Earth. Dengue fever, which I had the misfortune of contracting once in Mexico, was one of the most debilitating experiences of my life, and then there's bot fly, which lay their larvae inside an infected wound.

Several of my soldiers complained of itchy sores for weeks after our jungle exercise, and discovered to their horror that maggots an inch long had been growing underneath their scalps. Leishmaniasis, Chikungunya, West Nile virus, Zika, Ebola, monkeypox and histoplasmosis are all maladies to be found in tropical rainforests, and for many, all the more reason to stay away.

With so much potential for discomfort, disease and danger, it is no wonder that for those of us in the West, jungles and forests are often portrayed in literature, film and the media as *terra incognita,* or 'unknown land'. They can provide an interesting backdrop, artistic metaphor and a place of curiosity, but they are rarely shown as home. Forests are usually dismissed as dark, mysterious places harbouring too much of the unknown to be worth visiting, or incapable of sustaining human life – at least *civilized* life.

This notion of the forest as a dark, otherworldly space informs our psyche so much that when we speak of 'going into the woods', we are describing a confused, dangerous place where the outcome is uncertain. Other than being under the sea, jungles are the most alien of environments for those of us who grew up in towns and cities, and our culture and nomenclature reflect this modern relationship.

When we 'can't see the wood for the trees', it means we have lost perspective. Forests are where gangsters bury bodies and where horror films are set. Children are not to go into the woods for fear of getting lost. Carl Jung, the prominent psychiatrist, described the forest's fearful hold on us as 'primordial'. In the army we called the jungle 'The Green Hell'. Our culture is filled with warnings about the dangers that lurk in dark forests, from Hansel and Gretel, who are kidnapped by the child-eating witch, to Little Red Riding Hood who encounters the Big Bad Wolf.

The Amazon rainforest is the world's greatest single expanse of tropical terrestrial life, a rainforest the size of the United States, or as Wade Davis called it, 'a blanket of biological wealth as large as the face of the full moon'. And yet, Joseph Conrad described the jungle as less of a forest than a primaeval mob, a remnant of an era when vegetation rioted and consumed the world.'[1] A contemptuous view that had been perpetuated for centuries.

Even some explorers, who by their nature wish to be challenged by new environments, have come to hold a view of the jungle that borders on disdain. In the 1930s, the explorer-priest Gaspar de Pinell described his own journey to the Amazon as a land 'where tall trees covered with growths and funeral mosses create a crypt so saddening that to the traveller it appears like walking through a tunnel of ghosts and witches.'

Equally when the British explorer Thomas Whiffen travelled down the Rio Putumayo in Colombia, he described the forest as 'innately malevolent, a horrible, most evil disposed enemy'.

These colonialist views are shared by many, but it was not always so. Rather than seeing forests as something to be feared, early humans saw them as something entirely different: *home*. The place that provided the cradle for our species, and the means with which not only to survive, but thrive.

When sheep and cows were domesticated 11,000 years ago in the Near East, humans began to settle in increasingly larger groups, first along the banks of the Euphrates and Tigris, and then across Asia Minor, the Levant and Egypt; what we refer to now as the Fertile Crescent. Soon after, the first villages and towns sprung up in Mesopotamia, and across the world people started to plant crops such as wheat, barley and rice.

Around the same time, millet was being farmed in China along the banks of the Yellow River, and a thousand years later the potato was being cultivated in the Andes, as were bananas in Papua New Guinea. As humans began to congregate in ever larger groups, more fields were needed to grow crops to feed a burgeoning population. The Neolithic Revolution commenced, setting the groundwork for human civilization. These fields required space, and where forests were in the way, they were cut down.

By 5000 BCE, civilizations across various ecological zones – from fertile river valleys to steep mountainous regions – had mastered the art of plant domestication. Essential grains such as wheat, rice and maize, supplemented by millet, barley, oats and rye, have sustained human societies through the ages and were the bedrock of civilization. When humans acquired knowledge of agriculture, many ceased to view the forests as home.

The benefits of this were numerous. It allowed groups of people to settle in one place instead of continually searching out new lands, and to grow to a size where they were better able to defend themselves. It also enabled some of that group to turn their hand to tasks beyond mere survival, such as building permanent dwellings. Agriculture did not end the subordination of people to nature, but as farming techniques were honed over generations, they became less dependent on the whims of what the land would give them.

Humans could not control when a herd of wild animals would migrate, but to a much larger extent, at least, they could control the continued regeneration of their own livestock. Humanity emerged from nomadic and semi-nomadic lifestyles that were deeply in touch with the ecology of forests, into forms of society that were routine, settled and increasingly out of touch with the rhythm of the trees. The familiar woodland domain gradually gave way to that of the field. Nature became something to be mastered, a nuisance to be tamed, as opposed to a garden to be cultivated.

Without knowing it, mankind had walked into a trap. Once a society has grown to a certain population size by using agriculture, it can never again revert to a simpler way of obtaining food – not unless it wants most of its members to die.

Somewhere along this journey, as we carved furrows into the earth and cultivated our crops, our bond with trees began to fray. When early humans left behind the trees, they also left behind their innocence. Thus came the dawn of the Anthropocene, where man attempted to remove himself from nature and enslave it to his own designs. From that time forth, the world would never look the same.

The populations of settled people grew quicker than those of hunter-gatherers, and children – no longer burdens who needed to be carried and looked after on the move – could play a part in an agricultural society. Either directly, by working in the fields and tending livestock, or by watching their younger siblings as their parents worked. As villages became towns, and towns became walled cities, parents could stray further away from their children, and for longer periods.

The remarkable evolutionary success of grains such as wheat has led some thinkers like *Sapiens* author Yuval Noah Harari to speculate that mankind did not domesticate wheat, but instead wheat 'domesticated' us. It is a sentiment shared by Chris Thomas, author of *Inheritors of the Earth*:

> It is true that most individual crop plants and domesticated animals will die to feed us, yet humans ensure that some live to spawn subsequent generations – an arrangement that has allowed for our ecological success. It is also true that an oak tree sheds a thousand acorns to enable the growth of a single offspring. The rest go to feed the squirrels and birds. It is this sacrifice that allows the oak to reproduce. Likewise, maize, rice and wheat now cover more than a third of all the world's cultivated land. They are the most successful plant species to exist today. It is therefore entirely valid to think of them as having taken advantage of a gullible primate to sow, fertilize and ensure their survival. Who then, is manipulating who?[2]

Perhaps it was the grasses, in their ceaseless battle with the trees, who had laid the trap for us?

Oscar Wilde dismissed nature as 'a place where birds fly around uncooked' and suggested that 'if nature had been comfortable, mankind would never have invented architecture.' Of course, he was merely expressing the prevailing opinion of the Victorian era. Wilde enjoyed taunting urbanites who escaped to the countryside or went off grand-touring at the weekend. The 19th-century wit preferred to *escape* nature. By now, the distinction between nature and humanity was arguably at its furthest point. But it had been a long time coming.

During the Neolithic migrations, mankind stripped continent after continent of its megafauna in every region we entered, killing off the likes of the mammoth, giant ground sloth and countless other large herbivores.

And yet, while those early human activities had a huge impact on certain ecosystems resulting in many extinctions, it was the dawn of agriculture and its associated deforestation that sped things up rapidly and began to reshape the landscape beyond recognition. The first agricultural revolution marked a turning point leading to large-scale clearance of forests for farmland.

All over the world, forests were being cut down to make way for humans. In Ancient Egypt, the limited forests on the banks of the Nile vanished as soon as the Pharaohs began to build their temples. Timber was even brought in from neighbouring countries such as Lebanon and Greece. From the Bronze Age onwards, deforestation intensified across the Mediterranean and the Near East, driven by the need for agricultural land and resources. Plato lamented the loss of lush forests in Italy, and Strabo, writing in the 1st century BCE, complained how the once-verdant forests of Cyprus had been largely cleared to make way for fields of wheat and barley, the staples of the ancient world.

Even before the Romans arrived in Britain, the landscape was largely cultivated. It is astonishing that, even before the introduction of metal, British farmers did a thorough job of clearing the wildwoods. It is thought that by 2000 BCE, almost half of Britain's forests had been cleared. It is a powerful folkloric picture to imagine Roman soldiers arriving on Britain's shores to encounter vast and foreboding forests hiding unknown numbers of untamed, hostile Celts. But modern archaeological techniques have revealed an altogether different story. By the time Julius Caesar landed at Kent in 43 CE, England was, in many respects, already a 'green and pleasant land'.

While the Roman occupation certainly brought new roads, towns and governance, for a great majority of Britons, life continued much as it had for the past thousand years. Its inhabitants were for the most part not savage woodland Celts thirsting for Roman blood, but small rural communities practising a familiar pattern of settled agriculture.

Significant felling of England's forests represented a Bronze Age phenomenon, not a mediaeval one. Literally so, when we consider the vast troves of bronze axes unearthed in sites like the 2007 hoard in South Dorset. These axes would have comprised the primary instruments in the felling and hewing of England's primordial forests. The ancient oak circle known as Seahenge is testimony to this fact. When 3D imaging was used to study the oak stump in the centre of the circle, it revealed the blade strokes of 59 different bronze axes. The greening of this pleasant land was nearly complete long before Roman sandals ever touched our shores.

While much of England's forests had been cleared for agriculture by this point, Roman infrastructure accelerated the pace of deforestation. The expansion of Roman settlements required timber for construction and land for farming, reshaping the countryside in irreversible ways.

Yet, even as the Romans reshaped the land, certain trees retained their powerful symbolic value. The olive tree played a gentler role in Roman culture, representing peace, renewal and divine blessing. In Greek tradition, Athena gifted the olive tree to Athens as a symbol of peace and prosperity. This enduring association with peace is seen in Virgil's *Aeneid*, where Aeneas holds an olive branch to offer reconciliation. The idea of forgiveness and reconciliation is instantly conveyed by the phrase 'extend the olive branch'. This metaphor finds its roots in the Pax Romana, where Roman envoys would carry with them olive branches as signs of goodwill. Before the white flag became the battlefield symbol for truce, Roman soldiers would literally 'extend the olive branch' to achieve the same effect.

The olive harvest itself, requiring communal effort and cooperation, could serve as a metaphor for societal harmony. But while the olive tree became a symbol of peace, Roman expansion into Britain brought anything but, as forests were felled to support the growing empire. Trees were both revered and exploited: valued for their cultural significance in one region, while destroyed for the sake of progress in another.

It is not difficult to imagine the olive harvest acting as a time for warring tribes or factions within a tribe to set aside their differences and collaborate on the difficult work of harvesting the gifts of the olive tree. Yet, even as olive branches were extended, the underlying tensions between those who cultivated the land and those who sought to conquer it were never far beneath the surface.

Not all agricultural societies were unsustainable, of course. There are many examples of agrarian communities that managed to exist in harmony with nature. 'Forest gardening' is considered the earliest form of farming, and is still a practice widespread across India, Southeast Asia, Central Africa and South America.

In the Amazon, as we have recently discovered, people have been cultivating crops and regulating the landscape for millennia. An archaeological study published in 2020 confirmed that crops were being cultivated on man-made islands for at least 10,000 years. This makes the Amazon one of the earliest known locations for crop domestication anywhere on Earth, maybe even predating the Fertile Crescent.

When the first European explorers sent back accounts of cities, roads and farms on the banks of the Amazon River, they were met with disbelief. Gaspar de Carvajal, a Dominican missionary and chronicler who travelled with the conquistadors up the mighty river in 1542, wrote that one town 'stretched for 15 miles without any space from house-to-house which was a marvellous thing to behold.'

Even as recently as the 20th century, these unverified fables still held a fascination among geographers and explorers. It was tales such as these that convinced Percy Fawcett to set off on his ill-fated adventures to discover the Lost City of Z. Whether he found it or not is unknown. After 1925, Fawcett was never heard from again.

Fawcett was no 'kook', but a highly decorated and respected explorer, having received the Gold Medal of the Royal Geographical Society. Through his time mapping and exploring the Amazon, he had noticed recurring themes in the stories of lost cities and palaces and reasoned there must be some truth if all stories were so similar.[3] He found a document in the Rio de Janeiro Colonial Archive entitled 'Historical account of a large, hidden, and very ancient city, without inhabitants, discovered in the year 1753'. The account of a Portuguese soldier of fortune speaks of ruins of a spectacular city, complete with statues, a temple, stone arches, wide roads and hieroglyphics spattered throughout.

With the discovery in 1911 of Machu Picchu in the Peruvian Andes by the marvellously named Hiram Bingham III, the idea of lost

Amazonian cities may not have seemed so far-fetched. Fawcett's youngest son, Brian, described his intrepid father: 'True, he dreamed; but his dreams were built upon reason, and he was not the man to shirk the effort to turn theory into fact.'[4]

Accounts from the conquistadors were often dismissed as inventions for centuries, because no archaeological ruins were ever found. Some of this was based on the firmly held belief that no population of 'savages' could have possessed the skill or the inclination to build cities. That was the domain of the white man.

However, more recent research demonstrates that the old chroniclers were not mere fantasists. In the 15th century, large urban settlements did indeed exist on the banks of the great rivers of the Amazon. Their inhabitants were not hunters and gatherers, but sophisticated farmers who managed the soils, had permanent fields, and turned much of the forest into orchards.

Recent aerial surveys with LiDAR technology – infrared cameras that can capture images through the trees' canopy and show the lay of the land – now support Fawcett's speculations and the accounts of the chroniclers. Charles Clement of Brazil's National Institute of Amazonian Research says that 'few if any pristine landscapes remained in 1492', when the Europeans arrived in the Americas. 'Many present Amazon forests, while seemingly natural, are domesticated.'[5]

Anna Roosevelt, great-granddaughter of the 26th American president (and Amazonian explorer) Theodore Roosevelt, is an archaeologist at the University of Illinois. She went in search of these lost cities, only to discover a large flat island at the mouth of the Amazon, and massive earth mounds connected by causeways and canals. These, she believed, were towns that were home to tens of thousands of people, and evidence of a sophisticated system of trade that traversed South America.

Researchers from the University of Florida found similar structures a thousand kilometres away in the upper Xingu valley, with what appeared to be castles, bridges and dykes, all linked by straight roads, some 50 metres wide. Clark Erickson at the University of Pennsylvania described the engineering as being 'comparable in scale to the pyramids'. This careful cultivation kept Amazonian crops safe from flooding, while canals delivered water through a complex irrigation system when the atmosphere was dry. People alive at the time 'completely altered' a landscape that now lies 'empty and abandoned, save for the regrowth of forests'.[6]

What happened next must act as a cautionary tale when it comes to civilization.

The Spanish brought with them disease and illnesses that spread like wildfire among the indigenous population of the Americas, who had no immunity to the pathogens. It was a holocaust on an unimaginable scale, resulting in the deaths of some 20 million people[7] – a nightmarish 95 per cent of the native population. The towns and cities that had hitherto thrived were soon depopulated, and nature took over as the abandoned cities of gold became reclaimed by the rainforest.

The locals that survived, disappeared into the jungle, abandoning their farms and going back to the ways of the hunter-gatherer. It was their only chance of survival. What we now consider 'uncontacted tribes' are the remnant of those original refugees from disease. The memory of what the invaders brought lives long in their collective memory. Far from being 'uncontacted', they are instead scarred by human touch.

Chapter 7

The Folly of Man

———

Fancy cutting down all those beautiful trees . . . to make pulp for those bloody newspapers, and calling it civilization.

> – Winston Churchill, during a visit to
> Canada in 1929

When Rapa Nui, also known as Easter Island, was 'discovered' off the coast of Chile by Europeans on Easter Sunday in 1722, it was a barren landscape stripped of almost all its trees. The small number of inhabitants, just six or seven hundred, lived in a state of civil disorder and were thin and emaciated. Virtually no animals besides rats inhabited the island and the natives lacked seaworthy boats.

Understandably, the Europeans were mystified by the presence of great standing-stone statues, ten metres tall and weighing over eighty tons. Even more impressive were the abandoned statues that littered the floor at twice the size and weight. Who were these fallen gods? And how could such primitive people create and then move such enormous structures? The mystery of Easter Island stumped historians for centuries.

A tiny 63-square-mile patch of land more than a thousand miles from civilization, Rapa Nui saw humans make landfall on its blissful shores perhaps around 1,500 years ago. A small collective of Polynesians – maybe even just a single family – settled and began cultivating resources. Prior to their arrival, the place housed as many as 16 million trees, some reaching as high as 30 metres. The island was even home to an endemic forest of palms, perfect for constructing the new society's necessities, such as canoes, homes and even ropes (for shifting giant stone statues).

The people who would later become known as Easter Islanders had plentiful access to fuels and timber, using their seaworthy canoes to live off the fish and porpoise that swam in the coastal waters. They would go on to create steady communities with a complex societal structure, complete with a centralized government and religious authorities.

Historians think it was these communities that carved the enigmatic statues of Easter Island during their civilization's peak when the population was in the thousands. However, hauling the massive stones around the island on rudimentary wooden platforms and rope began to impact the island's ecosystem. Pollen analysis reveals a rapid decline in the island's tree population as deforestation took its toll.[1] Endemic palm species vanished, unable to reproduce due to overharvesting and a burgeoning population of rats, who, having arrived as stowaways with the first settlers, consumed their seeds and seedlings.

In the years after the disappearance of the palms, evidence from ancient garbage piles shows that the islanders began struggling for food. Without palm trunks to craft seaworthy canoes, they had no way to hunt for porpoise and were forced to turn to land-based farming, further clearing the forest to make arable land. Desperate for protein, the islanders took to consuming native birds, molluscs and even the dreaded rats. As migratory bird populations diminished, the forests teetered on the brink, deprived of vital pollinators and seed-dispersing birds.

With the forests gone, the islanders fell into trouble. Streams and drinking water dried up, crops failed due to harsh weather conditions, and topsoil eroded. Fires became a luxury without wood, and they took to burning the grasses as fuel. The gods they once revered seemed to have abandoned them and starvation set in, followed by anarchy. When Captain James Cook arrived, he found a miserable scene: the remnants of the population hungry, their canoes now nothing but fragments of driftwood. Even the statues were vandalized, as rival groups desecrated the gods that had forsaken them.

The Polynesians made a fatal error, killing trees to cultivate a society that would then worship statues, and in doing so destroying an island from which they had no escape. Hubris and greed played a part in their downfall, but the greatest tragedy is perhaps their perceived nobility in doing so. As the Canadian scientist and poet Hubert Reeves once said, 'Man is the most insane species. He worships an invisible God and destroys a visible Nature. Unaware that this Nature he's destroying is this God he's worshipping.'

Unsustainability has its roots in ignorance but is often driven by greed. The consequences of the agrarian revolutions were booming populations and unchecked growth. These fuelled the constant need for expansion, conquest, war and exploitation – some of the uglier aspects of the last one thousand years.

The Vikings' reputation for pillaging is well known, a society reliant on slavery and land ownership with most of the land being owned by chieftains. It was this mentality that led them to being some of the most brutal colonizers of the era. The Viking exploitation of Greenland, which lasted from the 10th to the 15th century, had a significant impact on the depletion of natural resources across the island. The Vikings were originally drawn to Greenland because of its extensive lands and potential farming opportunities.

However, their unsustainable agricultural practices and overexploitation of natural resources led to significant environmental degradation. They brought large numbers of cattle, sheep and goats to Greenland, which required extensive grazing land. Large areas were cleared to make way for pastures, peat was burned and the forested coastline was stripped bare. This deforestation led to soil erosion and ultimately ecosystem collapse. The last Norse colony gave up and either died off or went back to Scandinavia around 1420.

During the Middle Ages, an estimated 95 per cent of European houses were made using wood. It is no surprise that availability of timber profoundly influenced the growth and development of European cities. Forests in proximity to urban centres provided a steady supply of building materials, enabling rapid expansion of housing and city walls as the human footprint grew bigger. Timber-framed structures, often several storeys high, became emblematic of mediaeval European architecture.

As cities flourished, so did their industries. Blacksmiths, wheelwrights and carpenters thrived, using timber as their primary resource. This economic vitality created rising demand for timber extraction and trade. As the population grew and the demand for timber increased, countries needed to learn how to become more sustainable, and so coppicing was introduced.

Coppicing is a system of managed forestry where certain types of trees are selectively felled and allowed to resprout from their trunks to redevelop. The cycle allows for a continual supply of wood. It became common practice in places like England, where the ancient forests had been destroyed in prehistory. But even this was not sufficient to meet the demands of a burgeoning population and its insatiable demand for firewood and construction timber.

As the northern European countries chiselled away at their own natural resources, in the late 15th and early 16th century it was the turn of Spain and Portugal to take the initiative on where to get their own slice of the pie.

From the mid-16th century, this shortage became more acute, and the price of wood rose inexorably. As Lewis Dartnell explains in his book *Origins*, Europe was hitting 'peak wood': all suitable land was being used to grow food and the production of fuel could not be increased any further.[2]

The timber trade started to play a pivotal role in shaping economic dynamics and power struggles across the continent. The demand for wood, especially oak, soared due to its indispensable role in shipbuilding, a crucial aspect of maritime trade and military dominance during that era.

One notable example is the exploitation of timber resources in the Baltic Sea region. Stretching across present-day Scandinavia, northern Europe and the Baltic states, this area has vast forests with prized timber resources. European powers, including the Hanseatic League, penetrated deep into the Baltic hinterlands, establishing trading outposts and negotiating lucrative timber contracts with local rulers and forest communities.

The exploitation had profound consequences for the region's environment, economy and geopolitics, which not only altered the landscape but disrupted ecosystems and threatened biodiversity. Millions of cubic metres of forest were harvested annually during the peak of the trade. The Baltic timber trade also fuelled geopolitical rivalries and conflicts, as European powers vied for control over key trading routes and forest territories. Competing interests often led to diplomatic tensions, trade disputes and even military confrontations, shaping the course of European history.

There were more mouths than ever to feed, and nations needed to start looking beyond their own borders for resources. Things started to get competitive, and the age of Empire was about to begin.

The Iberian nations established great navies and began on a course of conquest and colonization on a scale the world had not seen before. While Spain's conquistadors set sail in galleons seeking new lands and plunder, similar expansionist drives were unfolding elsewhere. Portugal was staking claims in Africa and Asia, while other emerging European powers eyed territories in the Americas and beyond. This era marked a profound shift in global dynamics, driven by the relentless quest for resources and dominance. A region that I know well personally emerged as a focal point of Spanish ambition.

Nestled along the azure waters of the Caribbean Sea and the Gulf of Mexico, the Yucatán Peninsula has been a place that has drawn me back countless times over the last 20 years. I used to live in the city of Merida, home to the oldest cathedral in the Americas. Once I even set off on a journey to walk from Mexico to Colombia, starting on the very coastline where the notorious Hernán Cortés and his conquistadors made their landing in 1519, adorned in gleaming armour, to overthrow the Aztec Empire and establish dominion over the New World.

The Peninsula's ancient limestone bedrock gives rise to a landscape of enchanting cenotes (*water-filled sinkholes*), lush forests and sprawling savannahs. But the most fascinating aspect of this land is its rich history of Maya civilizations that thrived here well over a thousand years ago. Their legacy is evident in towering ancient ruins, pyramids and the myths, legends, stories and beliefs they left behind.

Much of this natural wonder was lost because of colonization. I met Professor Iván Batún, of the Universidad de Oriente in Yucatán, on the campus near Valladolid. We shook hands under the shade of a blooming red Delonix regia tree.

'There is a good example of how colonization depleted the natural environment, and that is beekeeping. Bees play a very important role in pollinating plants and helping trees to flourish,' he told me. This period of rapid colonization and ecosystem destruction saw unintended and unexpected environmental consequences.

To understand what happened with the bees, we must place ourselves in the Yucatán in Central America in the mid-16th century, which was then under the fierce dominion of Diego de Landa, a Spanish Franciscan priest who was ordained as the Bishop of Yucatán. A fervent missionary from a noble Spanish family, Landa had asked to be sent to the New World to convert the native Mayas to Christianity. Unfortunately, his impact was to be as controversial as it was long-lasting.

Landa spent time learning the Mayan language, codices and culture. It was precisely because of his interest in the Mayan language that he was shown their closely guarded books, which documented their history, beliefs and astronomy. Unsurprisingly, this body of work did not align with his Christian principles, and he determined that there was 'nothing in which there was not to be seen as superstition and lies of the devil'. Full of religious fervour, in 1561 he ordered these books to be burned, decimating a deeply significant body of literature and a precious part of Maya cultural legacy.

His own legacy did not stop there. Using tactics from the Spanish Inquisition, he was responsible for the torture of 4,500 people, and nearly 200 deaths. Even by the standards of the day, his methods were considered exceptionally brutal, and he was remanded in Spain to face trial for his actions. Back in his native land, he was condemned by the Council of the Indies, but later exonerated by crown authorities and returned to the Yucatán, where he continued to exact an oppressive regime before eventually dying in 1579.

For the Mayan population, the negative impact of his presence was significant. However, it was not only the European *people* who were to have such a devastating effect on the Yucatán. Their religion was equally catastrophic.

A few years earlier, the Vatican had convened the Council of Trent as a response to the growing Protestant movement in Europe. Pope Paul III and his successors sought to clarify a few ecumenical matters that had far-reaching consequences.

As Landa walked around the Yucatán barefoot, preaching the Bible and torturing unbelievers, the Pope issued an edict ordering that all prayers must be conducted in the presence of a lit candle. Candles had been part of church services for a long time. However, this drive to formalize services, including sacraments and mass, called for the procurement of considerable quantities of candles.

'As Catholicism spread in Mexico and Latin America, so did the requirement for beeswax,' explained Professor Batún. 'The Spaniards started to demand ever more wax as part of the village "tribute" – or tax.'

Beekeeping had already been a major part of the Mayan indigenous agroforestry system for thousands of years. The native, stingless Melipona bees were famed for their delicious honey and even honoured in Maya religious ceremonies. But the new demand for candle wax far outstripped the bees ability to regenerate it. As more wax was harvested, the bees had nowhere to lay their eggs and could not reproduce fast enough. The bees began to die off.

As with all changes in the delicate balance of nature, the reduction of Melipona bees had ecological consequences. These native bees had specific pollination relationships with local plant and tree species, developed over a considerable period of time, and their absence disrupted this critical symbiosis. Some of the plant species reliant on Melipona bees for pollination started to experience reduced reproductive success and even local extinction, and the loss of the bees as pollinators affected the overall biodiversity of the region.

'Can you imagine? Candles and the Catholic Church messed up the entire system,' said Batún, shaking his head. This wake of destruction was not limited to beekeeping, though. It was only one of the unintended consequences of disrupting an ecosystem.

The Spanish were greedy for maize and corn to provide food for an ever-expanding empire and embarked on large-scale farming to accomplish this. In their efforts to harvest greater quantities of these cereal crops, they initiated a programme of mass deforestation. Over a period of centuries, the forests of the Yucatán were systematically cleared and the landscape forever changed to accommodate monoculture farming. Colonialism was the catalyst for one of the first large-scale, human-induced deforestations of the world.

The Yucatán was once a fertile region, rich in agricultural resources that had been developed over time through sophisticated forest-farming practices. Its resources also extended beyond agriculture to timber and minerals, which offered the potential to generate great wealth. As the Spanish Empire expanded and its wealth grew, so did the look of envy from the likes of England.

There is a lingering myth among Western ecologists, who often cite the Maya civilization alongside Easter Island as an example of how a culture becomes its own undoing through ecocide. Some scholars have argued that in building the great pyramids at Chichén Itzá, Uxmal and so on, the Mayas cut down all their forests, destroyed their soil and overexploited their resources leading to ruin. So that by the time the Spaniards arrived in the 16th century, the entire Yucatán was already decimated.

'Not so,' said Batún. 'All the stuff about the Mayas chopping down all the trees is untrue. They managed the forests using centuries of indigenous wisdom. They used slash and burn techniques to set and control fires, which helped the soil create nutrients. And they also let the soil rest every few years to regenerate.

'Here in the tropics, if you leave a piece of land for a few months, trees will start to grow quickly: a tree can grow to 20 metres or more in just a few years. The reason these cities were abandoned is simply because new ones were built elsewhere, or, like in Europe, they were attacked by neighbours and destroyed. In other cases, seasonal drought forced people to migrate. The Mayas never overexploited nature, they lived as part of it.'

'Milpa,' he told me, 'is a system they developed to grow different crops together to ensure soil consistency, for instance growing corn with beans and squash. This forest gardening did not have a huge impact on the environment, and it was a very efficient management system. It was colonization that changed everything; colonization brought monocultures and greed. When the Spanish wanted corn, they got rid of the old Milpa system and chopped down the forest to create these corn fields. We used to have seventeen species of corn, but then guess what happened, it became a monoculture and the land became tired. When soil gets depleted, trees cannot grow.'

As the Spanish Conquistadors carved their mark across the Americas, the geopolitical landscape of the early 16th century was dramatically shifting. The conquests of the New World and the subsequent establishment of Spanish dominance showcased the transformative power of empire and exploitation. However, as European nations began to vie for global influence, the intellectual and scientific advancements that would underpin future imperial endeavours were simultaneously taking shape. This period of exploration and conquest set the stage for new forms of knowledge and governance that would influence subsequent powers, particularly Britain.

As centuries passed, mankind's desire for dominion only grew. By the late 16th century, this control evolved into a system where chartered companies, such as the East India Company, were granted exclusive rights by the state to manage and exploit colonial trade. These companies, operating under state-backed monopolies, effectively became the primary agents of resource extraction and trade, marking a significant shift in how nations engaged with their colonies.

Few figures are more closely associated with the intellectual foundations that supported the rise of the British Empire than Francis Bacon. In addition to being an adept statesman and patron of the arts, Bacon was a man of science, or 'natural philosophy' as it was then known, and he played a key role in transforming the Western world's approach towards nature. Even today, half a millennium later, we are in many ways still governed by scientific modes of thinking that find their origins in Bacon's thought.

Bacon is perhaps most famous for developing the principles that would shape the modern scientific method, a mode of experimental observation that emphasizes the study of nature in a systematic and empirical way. Rather than viewing nature as a sacred entity, Bacon saw it as a domain to be understood, controlled and utilized to expand human knowledge and potential. This approach is evident in his work *Novum Organum*, where he outlined a revolutionary method for studying nature, advocating for an empirical approach that prioritized human mastery over the natural world.

While often depicted as a bold luminary of secular scientific investigation, the reality is that Bacon was an orthodox Christian, and he rooted much of his natural philosophy in his interpretation of Holy Scripture. He was known to quote scripture often, and included a great deal of biblical imagery in much of his work. In his book *The Religious Foundations of Francis Bacon's Thought*, the scholar Stephen McKnight argues against the idea that Bacon was some sort of dispassionate secular materialist, insisting to the contrary that he was a deeply religious man whose scientific theories were steeped in his understanding of the Christian faith.

McKnight and others point to tracts like Bacon's work of utopian fiction *New Atlantis* (1627), where he combines faith and revelation with his scientific views. Rather than turning away from Christianity, Bacon grounds his philosophy in the idea of Christian dominion over the natural world. There's that word again: *dominion*.

It is hard to understate how influential Bacon's ideas would become. Bacon almost single-handedly paved the way for a generation of Enlightenment thinkers to develop a more systematic and empirical approach to understanding nature and man's place within, or over it, as was often the case. The Scientific Revolution was, in no small way, the result of the broad-scale implementation of this approach, which itself was rooted in Bacon's scientific method.

This new way of thinking was the nail in the coffin for what remained of any European tendency to revere the natural world as something living and holy, with rights and mysteries all its own. Bacon turned nature into a resource to be utilized, something that could be chopped up and put into boxes, pinned onto boards or plucked into test tubes.

Of course, the scientific culture that was born in this era has produced a great many groundbreaking discoveries of enormous value to humanity. From the invention of penicillin to the eradication of polio, the scientific method produced modern, man-made miracles that are worthy of our gratitude and our praise. But there were costs as well: the great price that we paid for this conscious shift was the severing of a cord that connected humanity to the natural world. A connection that emphasized that we were a part *of* this world, not apart *from* it.

Of course, Bacon's view was not without its critics. The influential French philosopher Jean-Jacques Rousseau fiercely opposed the notion that nature existed solely for human exploitation. Instead, Rousseau championed a vision of harmony between humans and the natural world, arguing that early human societies lived in a state of balance with their environment – an idea later simplified into the concept of the 'noble savage'. Unlike Bacon, Rousseau's critique was grounded not in religious myth but in a speculative anthropology that envisioned a more primal connection to nature.

While Rousseau acknowledged the brutality of early human life, he believed that these societies were intimately connected to nature's rhythms in ways almost unrecognisable today, except in the practices of Earth's remaining indigenous populations. Although Francis Bacon and Jean-Jacques Rousseau lived in different eras, Rousseau's critique of Bacon's mechanistic view of nature reflects a broader philosophical divide that spanned generations.

In his famously controversial ode to white supremacy, Rudyard Kipling wrote:

> *Take up the White Man's burden –*
> *Send forth the best ye breed* [. . .]
> *To wait in heavy harness*
> *On fluttered folk and wild –*
> *Your new-caught, sullen peoples,*
> *Half devil and half child.*

Kipling's poem, written as an exhortation for Britain's American cousins to tighten their grip on the unruly Philippines, has long been considered emblematic of the worst instincts of imperialism and colonialism. It embodies what Kipling himself described as the Englishman's 'Divine Burden to reign God's Empire on Earth'. For the European colonialists, exercising 'dominion' over both the natural world and its inhabitants was a key aspect of their civilizing mission.

This is a surprising assertion from someone who described himself confusingly in 1908 as a 'God-fearing Christian atheist'. But the sentiment that Kipling conveys in this poem extends not merely to the 'new-caught, sullen peoples, half devil and half child', whom he saw as the natural subjects of Anglo-Saxon dominion, but also to the natural world from which these subjects sprung.

Colonial literature is rife with examples of indigenous peoples being depicted as emerging *from* or being assimilated *into* the natural environment in a way that the 'civilized' Christian citizen of the West was not. Nature, in all her wildness and mystery, was seen to have an organic, 'devilish' hold on such populations, in stark contrast with the chaste and upstanding colonial emissaries of white Christendom.

The reining-in of such populations was seen as a critical aspect of the broader project of exercising 'dominion' over the natural world.

In many indigenous cultures, forests were seen as sacred spaces, imbued with spiritual significance and home to a diversity of life that was respected and revered. The Spanish in South America encountered complex animist rituals that demonstrated a deep reverence for nature through the use of holistic plant medicines. However, these practices were quickly suppressed as pagan aberrations by the colonial regime, which sought to impose its own values and exploit the land for economic gain.

The hypocrisy is staggering. Plants have also provided a source of natural stimulation for thousands of years in the Western world. Some 5,000 years ago, in the foothills of the Eastern Himalayas, some inquisitive humans discovered that by brewing them in boiling water, the bitter leaves of a low-growing evergreen tree in the local hills could be made into a palatable and invigorating drink.

This discovery would be a significant one for mankind, resulting in tea becoming the second-most consumed liquid in the world after water. Across Southeast Asia, people had been chewing on tea leaves for millennia, but depending on who you ask, the Chinese, Burmese or Indians invented the popular brew in its current form around 3000 BCE. Its cultivation spread across the region as it gained popularity and by 1610, it had reached Europe.

Soon after tea was introduced to Europe, China sought to protect its dominance by banning the export of tea trees or their seeds. However, by the early 19th century, British adventurers, driven by the demands of empire, embarked on covert 'tea raids' from British India into China. Disguised as locals, they smuggled tea plants back across the mountains to cultivate their own plantations in Assam and Sikkim. In 1823, a new type of tea plant was discovered in the Indian rainforest, which was hybridized with the Chinese variant, giving birth to what we now recognize as black Assam tea – a product that would fuel the engines of the British Empire.

The success of this 'imperial brew' came at a cost. In the pursuit of global dominance, European colonialists consumed not just tea but also vast tracts of land, clearing forests at an alarming rate to make way for plantations and urban development. As tea plantations thrived, ancient forests disappeared. Today, there are no truly wild tea trees left, as they have been hybridized and cultivated for centuries.

While European powers saw nature as something to be exploited and tamed, indigenous peoples in many of the colonized regions had a fundamentally different worldview. In northern India, Mexico and other parts of the world, reverence for nature was central to indigenous life. Their pantheon of deities reflected this deep connection to the earth, with gods of maize, rain, sun and flora standing at the heart of their belief systems. Plants were not simply resources – they were sacred and essential to all aspects of life, from food and medicine to spirituality.

The colonialists dismissed these traditions, prioritizing rapid deforestation to cultivate monocrops like tea and sugarcane, or to build new cities. Long-term ecological balance was cast aside for short-term gains. The contrast between these worldviews – indigenous respect for the earth versus European exploitation – became one of the defining aspects of colonial expansion.

The British Empire's enlargement was built on more than just tea. As British cities swelled and the empire's needs grew, war and conquest became the engines of its expansion. To support these ventures, vast armadas of ships were constructed, requiring timber from forests across the globe. From the cedar forests of Lebanon to the bamboo groves of China, Britain harvested the wood necessary to fuel its military ambitions.

The British landscape is synonymous with rolling green fields. We already know that significant deforestation happened during the Bronze Age, and the Romans sped up deforestation too, but by the time the first Elizabeth became queen, England had felled most of its remaining ancient forests to build the navy that would secure its empire for four centuries.

Nestled within the rugged terrain of Dartmoor, Wistman's Wood stands as a reminder of a time when western Britain was covered in temperate rainforests. Today, as one of the last remaining fragments of rainforest in England, Wistman's Wood is a vital ecosystem in its own right, providing a sanctuary for a rich diversity of plant and animal life. The tiny biosphere harbours over 100 species of lichens and trees over 400 years old.

It is a rare survivor, a testament to the cultural and ecological richness that once defined England's woodlands. But oak trees like those in Wistman's Wood, with their sturdy trunks and sprawling branches, played a particularly crucial role in England's history of deforestation. Prized for their strength and durability, they were the preferred choice for shipbuilding during the age of empire.

Britain emerged as a dominant naval power, epitomized by the famous phrase 'Britannia rules the waves', but it came at a cost – that would be paid by Britain's forests.

Britain's naval supremacy owed much to the formidable oak tree. During the peak of Horatio Nelson's naval era in the late 1700s and early 1800s, constructing a single 100-gun ship of the line required approximately 4,000 oak trees or up to 40 hectares of forest. To put this into perspective, it equates to the exploitation of an expanse equivalent to 3,750 city blocks of densely packed oak forest. Remarkably, these ships, on average, served for only 12 years. That gives some indication of the vast scale of deforestation that must have occurred to clear the path for Britain's colonial ambitions – chopping, sawing and sanding these peaceful conservators of life into lifeless vehicles transporting colonizers across the globe.

Inevitably, as the demand for timber grew, so too did the pressure on England's forests. During the 17th and 18th centuries, the British Royal Navy was consuming an estimated 70,000 mature oak trees annually for shipbuilding. From an estimated land coverage of 15 per cent in 1086, England's forests and woods had dwindled to just 6 per cent by the start of the 20th century.

The emergence of industrialism brought with it unprecedented economic growth, technological innovation and social transformation.

However, it also ushered in an era marked by the relentless pursuit of profit, often at the expense of natural resources, including the world's forests. It is a system characterized by private ownership, rapid development and personal wealth.

While timber scarcity was not exclusively attributed to shipbuilding, the prospect of a shortage in shipbuilding timber was a preconceived outcome of pre-industrial naval expansion. The requirements of shipyards surpassed the capacity of woodlands to replenish, putting pressure on ecosystems already stretched thin by fuelwood production and other industries. In Europe, 19 million hectares of forest was cleared every decade between 1700 and 1850. If you are unsure what 19 million hectares looks like, take a look at Germany on a map. Imagine it covered in trees. Then imagine half of them gone. As empires continued to fight their wars, trees paid the price in their billions.

In 1760, Europe was on the cusp of an unprecedented transformation. The wheels of industry were turning, fuelled by coal, steam and the relentless pursuit of progress. Factories belched smoke into the sky. Cities swelled with the influx of rural migrants seeking their fortune in the burgeoning urban landscape. It was the dawn of the Industrial Revolution, a time of innovation and upheaval that would forever alter the course of human history – and leave an indelible mark on the natural world.

It was not a change that went unnoticed. In 1804, William Wordsworth lamented:

> *There was a time when meadow, grove,*
> *And stream,*
> *The earth, and every common sight,*

To me did seem
Apparelled in celestial light,
The glory and the freshness of a dream.
It is not now as it hath been of yore.
Turn wheresoe'er I may
By night or day
The things that I have seen I can now see no more.

Wordsworth's reflections came at a time when Europe passed a point of no return, and he raises parallels of his own childhood loss of innocence and a wider feeling that *'There hath passed away a glory from the earth.'* Begging the melancholic question:

Whither is fled the visionary gleam?
Where is it now the glory and the dream?

England's population more than doubled between 1700 and 1850, from around 5.5 million to over 12 million. With more mouths to feed and more bodies to shelter, the demand for timber skyrocketed. In my home town of Stoke-on-Trent, the forests were long gone, the coal mines and pottery kilns filled the air with black smoke, and men were lucky if they lived to 40.

William Blake wrote of 'dark Satanic Mills' in his poem *Milton*, which later became the hymn 'Jerusalem'. Industrial progress came at a price, and it was not pretty. By the late 19th century, almost all the ancient woodlands in England had been cleared, leaving behind barren landscapes and devastated ecosystems.

And it was not only in Britain where this was happening. Throughout the landscapes of Europe, forests were being flattened in the unabated march for progress. By the end of the 19th century, France's forest cover was reduced by a staggering 25 per cent.

Across the border in Germany, an era of intensive timber practices ensued, and with it bore the method of 'scientific forestry', where foresters would employ novel techniques that produced huge leaps in tree harvest yields, yet led to devastating habitat destruction and its consequences, such as soil erosion, increased flooding and loss of biodiversity.

At the same time, Britain's North American colonies were taking off, bringing with them goods manufactured in Britain and generating a trans-Atlantic demand. Money from the slave trade was being used to further Britain's infrastructure, such as building railways for coal-burning steam trains. The Industrial Revolution established burgeoning cities across the East Coast of the US, and Britain's insatiable demand for timber meant a death sentence for many American trees. Old-growth forests that once stretched across the continent were reduced to stumps. Communities of towering Pacific cedars and Douglas fir were diminished, leaving a gaping scar of a wasteland in its wake.

These effects were echoed all across the planet throughout the 19th century, with entire forests falling in Brazil, Indonesia and the Congo Basin as industrial logging, intensive farming practices and the need to house an exponentially growing population placed a burdensome toll on the planet's oldest and most intricate ecosystems.

War and conflict have had devastating consequences for the environment throughout the ages. Samson burned crops, vineyards and olive trees belonging to the Philistines, and Genghis Khan poisoned the water supply of mediaeval Baghdad. Half a million acres of French forest was lost to the fierce bombardments of the First World War. By 1918, after four years of brutal warfare and under the strain of a German naval blockade in the North Sea, England's forest cover had plummeted to just 4.5 per cent.

In the late 19th century, much of the wood required to fuel British industry came from the Baltic states when pine trees were plentiful, but during the war, the only way to survive was to cut down what was left of an already dwindling resource. Ancient oak and yew trees, some over a thousand years old, were felled to support the war effort, while new shipbuilding and trench systems demanded even more wood.

Returning soldiers, many suffering from shell shock, came home to landscapes stripped of forests and replaced by barren stumps. Areas once rich in oak, ash and thorn were left bare. Timber became a vital resource for rebuilding, but much of what remained was quickly replaced by fast-growing, non-native species like Sitka spruce and Douglas pine. These were planted in easy-to-manage rows, transforming the once diverse and thriving woodland into a landscape of uniform, non-native trees, permanently altering the natural environment.

The 20th century saw some of the greatest destruction of all. The atomic bomb destroyed entire ecosystems alongside its human victims in Japan during the Second World War and in the Pacific tests. During the Vietnam War, Agent Orange – a deadly concoction of chemicals designed to kill grass and trees and render the soil infertile – was sprayed from the air all over the lush jungles so as to lay siege to the hidden combatants. Eighty per cent of the U Minh Forest was destroyed by napalm bombardment. The legacy of these and other wars still scars landscapes around the world to this day.

Sadly, it seems we have still not learned our lesson, as I found out in 2023 when I travelled to war-torn Ukraine.

The limestone slopes, grasslands and birch woodlands of Kamianska Sich National Nature Park were Serhiy Skoryk's happy place. He would sometimes wander into the forests and stand in the near silence, tuning his ears to the subtle sounds of wildlife scurrying on the leafy ground and birds calling out to each other from above his head. He would smell the heady aroma of the steppe grass and amble along the banks of the Dnipro River, looking for signs of deer and monitoring the rewilding efforts. As the park director of the youngest national park in Ukraine, it was his job to protect the 12,000 hectares that was home to over 90 species of rare animals. He was a proud man content in his work.

On 24 February 2022, everything changed. Russia invaded Ukraine, commencing the largest attack on a European country since the Second World War. Serhiy was determined that no matter what, he would protect his country and he would protect his park. On the second day of the war, 25 February, Serhiy was captured by the invading Russian forces.

'They took me and some other men from the village for interrogation,' Serhiy told me as we drove towards the front line. I could hear the dull thud of mortars and artillery crashing somewhere in the distance. 'They beat and tortured us and locked us in a cold basement for many days. There were bodies everywhere. I thought they would kill us, and then one day they ordered the men to go outside and dig a hole. They were going to execute us all. So I decided to take my chances and run when they weren't looking.'

In an act of utter desperation, Serhiy dived into the half-frozen Dnipro River and swam for his life, almost two miles to the temporary safety of the western bank, where he was rescued by Ukrainian soldiers. It was a miracle he did not freeze to death. The fate of his colleagues was unknown. Serhiy was just one of many environmentalists caught up in this terrible war.

In the wake of conflict, the scars on the land often last the longest. In November 2023, eighteen months into the war on Ukraine, I visited the country to document the devastating environmental impact of the Russian invasion on biodiversity and plants, and to meet the resilient people caught in the crossfire.

Despite the dangers from drone strikes and Russian artillery, Serhiy drives into the park almost every day and while some of it has been cleared, most of the landscape is contaminated with explosive munitions. 'De-mining work will take years,' he tells me.

Kamianska Sich was once a sanctuary for diverse wildlife. Now it has become a silent victim of the collateral damage of war. A graveyard with only shreds of hope from the resilient people like Serhiy who remain determined to bring it back to life. His story is one of many that I heard of Ukrainians fighting on the front line, not only to save their people, but to save their environment and protect their nature.

'Nothing was as shocking as when we woke up to hear the news about the dam,' Serhiy told me. 'I didn't believe it at first.'

In the early morning of 6 June 2023, two explosions hit the Kakhovka Dam on the Dnipro River in southern Ukraine, releasing 18 cubic kilometres of water over 4 days and flooding over 620 square kilometres. Eighty settlements were hit by the surge of water, resulting in numerous casualties and leaving many missing. Around one million people lost access to drinking water.

The dam's destruction submerged villages and farmland, and around 150 tons of oil housed within the hydroelectric plant within the dam were swept into the water. Floodwaters also rushed into the Lower Dnipro National Park. 'There are catastrophic consequences for the environment,' Ukrainian environment minister Ruslan Strilets told reporters. 'For some of our ecosystems, we have lost them forever.'

We also travelled to the Sviati Hory National Park in the Donbas region. Before the war, up to 91 per cent of the park was forested. It was a place of majestic beauty known as the Holy Mountain National Park. Oaks – some over 500 years old – stood on the left bank of a river, surrounded by ash, lime, maple and pine. It was home to nearly a thousand unique, red-listed Cretaceous pines. One-third of the species of Ukraine were in the intricate ecosystem found within the park boundaries.

The statistics spoke volumes: an estimated 80 per cent of the park's biodiversity had been directly affected by the war. Over 5,000 hectares of forests were decimated early in the conflict. And yet, forest rangers, scientists and conservationists continued to do their important work.

I spoke with Ukrainian environmental scientist Kateryna Polyanska, who is continuing her work, against the odds, and is taking a stand for nature. I walked through the park, careful not to step off the path for fear of unexploded mines. Everywhere I looked was evidence of artillery strikes, craters and the aftermath of savage fires. Where towering trees once stood, they had been replaced by burned black earth. It was as if every snow-ridden branch screamed silently at me in pain, telling stories of destruction, displacement and struggle for survival.

It is the first war in history where the true cost of conflict to nature, every crime against nature, is being fully recorded, evidenced and added up. At the time of writing, it is estimated to be 2.4 trillion hryvnias in environmental damage according to the government. That is almost £45 billion![3]

'Hundreds of thousands of wild animals have been killed by explosions and mines,' Kateryna told me. 'We can try to put a monetary cost onto this to help the restoration effort, but really, nothing can put right the damage that has been done.' Putting herself on the front line, she is visiting Ukraine's national parks to record the destruction done by the war for a case now known as ecocide.

Serhiy wandered past the remains of a burned-out tank. He sighed. 'It's not just the trees and animals we are fighting for; it's our identity, our connection to the land,' he said. 'This war has tested us in ways we never imagined. It'll be a long time before any of this environment is restored. The park's infrastructure; vehicles, computers and equipment were stolen or destroyed. The impact on nature here is more immense than I ever could have imagined.'

In the bid to tell the human stories of war, the consequences on nature often get forgotten.

As horrific as the scars of war were on the landscape of Ukraine, as I drove away from the front line I could not help feeling that it was only a tiny part of the picture. Out of the window I looked upon what felt like endless wheat fields, which continued for hours. Forests still cover a significant portion of Ukraine and there are efforts to protect national parks, but Ukraine is also the provider of around a third of the world's grain. Even two years after the war broke out, the country was still exporting more than 5 million tons of grain a month.

Turning over the landscape to a monocrop has seen deforestation on an unprecedented scale in the last 25 years. No trees, disappearing insects, wildlife and the depletion of the soil have transformed the country into a dull picture indeed. You might say that this is not directly linked to the ongoing war, but the effects of the timber industry and industrial farming mean that the end state is the same.

Greed, fear and an overly competitive economic system all drive an unsustainable assault on nature, in an attempt to fuel a seemingly insatiable need to feed nations as cheaply and efficiently as possible. It is a race to the bottom and yet it is a story that is playing out all over the world at this very moment.

Chapter 8

Apocalypse Now

———

*But man is a part of nature, and his war against nature is inevitably
a war against himself.*

— Rachel Carson, during an interview for CBS

The air was thin and crisp as I stood at the summit of Mount
Nyiragongo gazing into the glowing precipice. My heart raced, partly
from the challenging ascent and partly from the awe-inspiring view
that reached out all around; the vast expanse of the Virunga National
Park – a refuge for some of the world's most remarkable wildlife, and
a battleground for conservation in the heart of Africa.

Mount Nyiragongo is an active stratovolcano, its slopes cloaked in
dense rainforest. The hike to the summit took me across a landscape
of towering trees and lush vegetation, through to open skies, over old
lava flows and past steaming volcanic fissures. The reward for the hike
was great – a vast, smouldering cauldron of molten red fire: the world's
largest lava lake.

I was travelling with conservationists and park rangers, all
working tirelessly to protect the park and its inhabitants. They
operate on the front lines in dangerous conditions. Only a few weeks

before my arrival, several rangers were killed in a gun battle with rebel fighters – 230 rangers have been killed here in the last two decades. Poaching here is rife, illegal logging funds militia groups, and the edges of the park are under constant encroachment by local farmers wanting extra land.

Virunga is a UNESCO World Heritage Site and is recognized as having the most diverse habitats of any park in Africa, acting as a sanctuary for some of the most endangered species on the planet.

In the 1970s, the Greater Virunga Landscape, spanning parks in the Democratic Republic of the Congo (DRC), Rwanda and Uganda, boasted the world's largest population of large mammals, among them its famed mountain gorillas and around 8,000 elephants, including the rare and critically endangered forest elephant.

However, wildlife began to decline due to conflict in the 1990s, sparked by the Rwandan genocide and subsequent Congo Wars. Many elephants fled to safer areas, while others fell victim to poachers seeking ivory tusks to sell for large profits. In recent years, the elephant population in Virunga dwindled to just 120 in total. However, in 2020, 500 elephants returned to the park as it provided the final refuge in the region.

Along the park's borders with Rwanda and Uganda, vast swathes of forest had been cleared leaving behind a scarred and barren landscape. I could see from my vantage point the contrast between the lush interior of the park and the stark deforested areas, which showed the challenges facing this fragile ecosystem.

For these rare elephants, and for most endangered animals around the world, it is the fragmentation of their habitats that is the largest threat to their survival. As forests are cleared for logging, agriculture and settlements, vital migration routes for wildlife are disrupted, increasing the likelihood of human–wildlife conflicts. In some cases, populations become isolated, resulting in reduced genetic diversity.

Their once vast habitat's quality and access to food become diminished, which increases the vulnerability of species to extinction, particularly in the face of disease outbreaks and the impact of climate change. Preventing the rampant destruction of the forests of Virunga means protecting the future of over 218 mammal species, 706 bird species, 109 reptile species and 78 amphibian species that live within the park's boundaries.

But in the case of Virunga, it is a constant struggle that is only getting worse. In recent times, conflict has driven huge numbers of refugees into the park. In 2023, tens of thousands of Congolese fled from clashes between rebels and the military and resorted to cutting down trees for firewood and charcoal to survive, often paying fees to militia groups to access Africa's oldest national park.

Over just four months, 2,382 acres (964 hectares) of forest was lost, according to forest monitoring platform Global Forest Watch. Twenty per cent of this loss has been around informal refugee camps located near the Nyiragongo volcano. Yet another example of conflicts having a devastating impact on nature. And despite efforts by park rangers and conservation organizations to combat poaching, the lucrative nature of wildlife trafficking also continues to undermine conservation efforts.

Over the border in Uganda, I encountered acres upon acres of burned woodlands, scenes of absolute decimation. 'It's for charcoal,' explained Boston, my Congolese guide. 'These people, they don't see why they shouldn't just take what is theirs. It's just wood, after all.' Charcoal accounts for up to 90 per cent of Africa's primary energy consumption needs, according to a 2018 report by the UN Food and

Agriculture Organization.[1] It serves as the main source of energy for cooking and heating water in Uganda due to its affordability and accessibility across income levels. But it is highly inefficient.

It is a complex interplay between socioeconomic factors, cultural practices and environmental conservation. 'They don't have affordable alternatives to charcoal or other ways for charcoal workers to make a living,' said Boston. Commercial charcoal making has recently been banned in northern Uganda, although it is clear the law does not have much of an effect in these rural areas.

According to a report by the World Wide Fund for Nature (WWF), rampant deforestation could lead to the loss of forest landscapes equivalent to an area more than twice the size of Texas over the next 15 years. Without intervention, 11 of the world's most crucial forest habitats, including those of orangutans, tigers and elephants, are projected to suffer over 80 per cent of global forest loss by 2030. One million plant and animal species are at risk of extinction,[2] with trees among the most vulnerable. Over a third of the world's tree species face extinction, accounting for more than a quarter of the listings on the International Union for Conservation of Nature's (IUCN) Red List. The number of threatened tree species is more than twice that of birds, mammals, reptiles and amphibians combined.[3]

And it is not just flora and fauna that are set to be affected. The biodiversity of these regions is critical for human wellbeing. These ecosystems regulate climate, purify air and water, ensure food security, pollination, and natural pest control. These species are part of a delicate food chain and once these ecosystems are damaged, it takes a long time for them to recover, if they ever do.

In 1979, an experiment was conducted in Brazil to ascertain the impact of forest fragmentation on species biodiversity. At that time it was becoming clear that rampant deforestation in the Amazon was having a huge impact on the distribution of wildlife. Deforestation would continue regardless, but the WWF wanted to at least see if there was a way to mitigate its impact.

Since landowners were required to leave 50 per cent of their land forested (allowing the other half to be cut down for cattle grazing), it was suggested that the clear cutting was done in one region to leave a number of 'islands' intact of varying sizes, in a bid to ascertain how species would react. There were five plots measuring one hectare, four plots of ten hectares and two larger ones of a hundred hectares.

Over the next 25 years the results came in and proved conclusive: large intact acres are very important, the larger the better. The small plots were basically useless; trees died, there was too much undergrowth and not nearly enough wildlife.

Many species, including the black spider monkey, need continuous tree cover and left to take up residency in nearby, continuous forest. Howler monkeys, on the other hand, seemed happy to stay put. The departure of one species can often mean the departure of another, as many rely on symbiotic relationships.

Trees once fostered an incredibly rich environment for biodiversity, laying the foundation for life on Earth, including our own. Now, in return, we are dismantling this very biodiversity. If we want to save wildlife, we need to save the trees.

Travel has allowed me to witness the impact of climate change and environmental destruction first-hand. From home, all of these issues can seem far away. But it is the people I have met in the more remote parts of the world that are suffering as a consequence of our greedy attitude to resources, whose stories are the hardest to hear.

I have seen landslides destroy villages in the Himalayas due to soil erosion where trees have been removed; illegal logging devastating communities in Central America; industrialized agriculture in the Amazon; illegal mining across Africa and monocultures in Asia. It is often the poorest countries who feel the effects of environmental change the most. But it is not those who are simply trying to feed their families that are most at fault.

The timber industry remains a primary driver of deforestation, particularly in tropical regions. While much timber is grown legally in forestry plantations, which can often be done sustainably, there are many areas where old-growth forests are cut down for furniture or poorly managed forests are clearcut, degrading the soil and leaving vast swathes of land barren and susceptible to erosion.

Around the world there are 25 million acres of fast-wood plantations, also known as commercially planted forests.[4] That is not to say they are all bad. After all, intensive production is necessary to satisfy global demand and it is better than chopping down natural forests. However, timber plantations do have detrimental impacts on the environment. Timber buildings can often be less harmful to the planet than other materials like concrete, yet it is a delicate balancing act of responsible sourcing from sustainable, commercial logging operations. Sadly a lot of timber does not come from legal or sustainable operations. Much of what ends up in our timber yards to make doors and desks is not what it claims to be.

Illegal logging is the world's 'most lucrative environmental crime', according to the United Nations, bringing in around $152 billion annually. Based on estimates, it constitutes approximately 10 per cent of the worldwide production and trade in forest goods and encompasses 40–50 per cent of all logging activities in some of the most critical forests on the planet. It is an industry that perpetuates corruption, jeopardizes livelihoods and incites social conflict.

Mining, particularly for gold and coal, is another major cause of deforestation. Over the past two decades, it has led to the loss of an estimated 6,877 square kilometres (4,273 square miles) of forest.[5] That is equivalent to an area the size of the state of Massachusetts.

Most of the mining-related deforestation that has occurred in the last decade has been in South America and Indonesia, but there have been sizable operations in Russia and Canada too. Protected areas are often downsized to accommodate mining, impacting millions of hectares of forest globally. Mining operations attract workers, leading to growing settlements and further forest clearance as they build roads to expand. But it is often not the miners themselves who are to blame. Yet again, the ever-increasing demand from developed countries is the driving force.

Mineral-consuming nations, particularly China, the US and European countries, while not primary miners themselves, import crucial minerals for their industries. 'Metals are important for the development of human civilization and the life we live today. But the footprint left by the extraction of these commodities has a heavy toll on ecosystems,' said the WWF mining global lead, Tobias Kind-Rieper.

Our smartphones contain some of the rarest elements on Earth. Metals like gold, iron, copper, aluminium, and the less familiar praseodymium and dysprosium are some of the elemental components contributing to the vibrant colours of phone displays, the smartphone speakers, and the durability of glass screens. Recent estimates indicate we will run out of some rare earths in the next 20–50 years.[6]

Cobalt is a vital component in lithium-ion batteries powering smartphones, laptops, electric vehicles and solar panels. The DRC accounts for 60 per cent of global cobalt production, much of it extracted from 'artisanal' mines, many of which are complicit in brutally long hours, hazardous conditions and low pay, often using child labour.

The gold-mining operations that support the tech industry contribute significantly to deforestation in the Amazon rainforest. Gold, essential for smartphone connectors and wires, produces waste rich in cyanide and mercury, posing serious threats to water and fish quality, and consequently, human health.

It comes back to over-consumption. We must understand the true cost of what we purchase and put systems in place to stop trees paying the ultimate price for our greed. The most eco-friendly smartphone is . . . the one you already have.

It seems we are now reaching a critical juncture in our history. Over the past century, the world has lost as much forest as was cleared in the previous 9,000 years combined. This deforestation is driven by a relentless pursuit of profit, a hallmark of industrial capitalism, concentrating wealth in the hands of a few. It is an unsustainable economic model and our collective survival on the planet almost certainly depends upon moving beyond it.[7]

Capitalism relies on growth. It relies on us requiring more commodities, more resources, more labour. The problem is the world's resources are not endless, they are finite. By defining natural resources purely based on their economic worth to humans, we alienate the natural world and ourselves. Ben Rawlence concludes in his book, *The Treeline*: 'Our very gaze has become a product. Our attention is directed away from the biosphere that sustains us, and this alienation makes us to varying degrees blind, deaf and dumb. Looking at the long history of our co-evolution with the forest, the human break with nature is an eye blink.'[8]

We are where we are because of decisions made generations ago. We did not choose this current economic system from a list of available options – capitalism was arguably the inevitable system to evolve out of a long procedural disconnect from nature. When material greed is lauded, land ownership is viewed as the ultimate status symbol, and the balance of power hangs purely by the fictional idea of monetary value, this is the position we find ourselves in.

The Western capitalist model demands resources for fast economic gain, many of which are found in the tropical regions; home to the oldest rainforests with the most diverse ecosystems on Earth. That is not to say other economic systems are any better. Until its collapse in 1991, the USSR created 1.5 times as much pollution as the US for every 1 unit of GDP. In North Korea, people are forced to hack away at forests for fuel, and Cuba also has problems with deforestation and loss of biodiversity.

Either way, our current global system is broken. Deforestation ranks as the second-largest contributor to human-caused greenhouse gas emissions after fossil fuel emissions, and stands as the main cause of biodiversity decline on land. If our capitalist system continues to discourage individuals from paying the monetary cost for a sustainable relationship with nature now, it will be humanity that pays the price later.

As global populations rapidly increase, the demand for resources is ever more pressing, especially global food production, which is the main driver of global deforestation. Sixty per cent of tropical deforestation stems from the production of beef, soybeans and palm oil. Brazil's significant contribution is primarily fuelled by its domestic beef demand, while in China, the main driver is the demand for oilseeds, a combination of soy from Latin America and palm oil from Indonesia and Malaysia. Most of this is not for domestic consumption, but for the use of the developed nations.

The system is designed for profit, not sustainability. We ship apples from Chile to Europe, we cut down old-growth forests in Canada to make 'biofuel' pellets to burn as green alternatives to coal in England, and bananas grown in Brazil get sent to Thailand to be packaged before being shipped to France. We use pesticides on our large-scale monoculture crops, permanently degrading soil that was once biodiverse forests, disrupting ecosystems and damaging water quality.

Borneo, the third-largest island in the world, is somewhere that captivates the imagination. A place where ancient rainforests used to teem with life, mist-shrouded mountains rise tall and fascinating cultures once thrived together. It is a vision of Borneo that, half a century ago, was easy to find. Today, only small pockets of its precious virgin rainforest remain. Borneo's primary rainforest has shrunk by over 30 per cent in the last 40 years. That means Borneo is experiencing deforestation at a pace twice as fast as other tropical rainforests worldwide.

I visited in the summer of 2022 to witness this destruction first-hand for a TV series I was filming for Channel 4. Setting off from the city, my route took me first through the vast palm oil plantations that now cover most of the country. 'This was all primary rainforest not too long ago,' said Dean, my local guide. 'Most of it is smallholdings owned by local people, they are just trying to make money for their families.'

Left **Methuselah Tree, California (possibly)** – This 4,800-year-old bristlecone pine is one of the oldest living trees on Earth. Its location remains a secret.

Below **A Depiction of Calamophyton, Early Trees** – Among the first trees to emerge, these Devonian pioneers stood 2–3 meters tall over 400 million years ago.

Above **Sugi Tree Circles, Japan** –
An experimental forest planted in
concentric circles with varying spacing,
showing how tree growth responds to
density and sunlight.

Left **Crown Shyness** – A natural
phenomenon in which tree canopies
avoid touching, creating visible gaps.
Scientists believe this helps prevent
the spread of disease and insects
while maximizing light.

Below ***Australopithecus Afarensis* in
the Forest** – An artistic interpretation
of early hominins navigating ancient
woodlands 3–4 million years ago.

Adam's Tree, Iraq – In a park near Basra, this tree is said to mark the site of the biblical Garden of Eden.

Yggdrasil, Norse Mythology – The mythological World Tree that connects heaven, earth, and the underworld in ancient Norse legends.

Green Man, Rosslyn Chapel – An ancient symbol carved in stone, representing the cycle of life, death, and renewal, deeply connected to nature and spirituality.

Left **Christ Depicted as the True Vine, Rosslyn Chapel** – Early Christianity took much inspiration from earlier pagan and animist religions. Jesus himself preached harmony with the natural world.

Below **Moai, Easter Island** – These iconic statues stand in a landscape where deforestation contributed to the collapse of an ancient society.

Far below **The Author at Yaxha Pyramids, Guatemala** – Ancient Mayan civilization flourished throughout the forests of Central America. The ruins of many lost cities were rediscovered after the Spanish Colonial era.

Above **HMS Victory, Battle of Trafalgar, 1805** – Nelson's flagship was built using around 6,000 trees, most of them oak.

Left **Amazon Deforestation** – Wildfires and industry-driven destruction are releasing greenhouse gases faster than the surviving trees can absorb, with much of the devastation driven by the global demand for beef products.

Below **Hanging Gardens of Babylon** – An ancient engineering wonder that brought lush greenery to the arid landscape of Mesopotamia.

Left **John Muir and Theodore Roosevelt, Mariposa Grove, 1903** – Standing at the Grizzly Giant, these two men helped ignite the U.S. conservation movement.

Left **Activist Julia Hill and Luna** – Living 180 feet up, Hill spent two years in this Giant Redwood to save it from destruction.

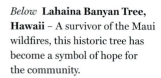

Below **Lahaina Banyan Tree, Hawaii** – A survivor of the Maui wildfires, this historic tree has become a symbol of hope for the community.

Benki Piyãko, Acre, Brazil – Chief and shaman of the Asháninka community, who has
planted millions of trees in a bid to reforest the Amazon. His mantra 'We are the Trees'
reflects ancient indigenous wisdom that we must respect the forest.

Above **Sycamore Gap Tree, UK** – The iconic tree near Hadrian's Wall, whose illegal felling sparked a global outcry of grief.

Left **The Author, Staffordshire, UK** – Beside the chestnut tree he planted in his parents' garden 30 years ago.

Palm oil in Borneo is a controversial and complicated issue. Look at the package for any product you have in your home, from shampoo to instant noodles to chocolate bars and it is more than likely it contains palm oil. Hailed as 'green gold', it has emerged as a cornerstone of Borneo's economy, contributing substantially to national GDP and providing employment opportunities.

On paper, it is brilliant. Palm oil is the most efficient vegetable oil due to the remarkable productivity of oil palm trees. In comparison to other oils, such as sunflower oil, it requires significantly less land. While one hectare of land yields 0.7 tonnes of sunflower oil, it can produce 3.8 tonnes of palm oil. The problem is, the land it grows on is some of the last remaining primary rainforests in the world. And once these ancient ecosystems are chopped down, it is very difficult, if not impossible, to get them back.

The island of Borneo is responsible for 85 per cent of the manufacturing of the world's palm oil, almost all of it exported. Even plant-based lifestyles are not totally free from a diet of deforestation. Fortunately, in 2014 the Roundtable on Sustainable Palm Oil (RSPO) was developed, supported by the WWF. This created a visible certification on products using only sustainably sourced palm oil, making it easier for consumers to make informed choices.[9]

While some local communities in Borneo can make money off smallholdings of palm oil trees, it is not always the case. The destruction of these ecosystems spells disaster for indigenous communities, who rely on these forests for their livelihoods and cultural heritage. Displacement, loss of access to traditional resources and conflicts over land rights are just some of the challenges faced by indigenous peoples as deforestation increases in Borneo, and all over the world. Despite assurances from the palm oil industry regarding sustainability initiatives and certification schemes, the reality on the ground paints a far bleaker picture. A plantation has about as much biodiversity and stores as much carbon as a single-species wheat field.

On my journey, I was in search of orangutans, the creatures who have paid the biggest price for palm oil. 'Orangutan is Malay for "person of the forest",' Dean told me. 'They share 97 per cent of their DNA with us.' In the last 60 years, deforestation and poaching have halved the population, with only around 100,000 animals left. These single-species plantations might look green and lush, but orangutans need to eat hundreds of different fruits to survive, including figs, bananas and durian, making it impossible for them to live here.

We visited Bukit Piton. Covering nearly 30,000 acres, equivalent to roughly twice the size of Manhattan, this area had undergone extensive logging since the 1980s. Subsequent fires further devastated the degraded forest. By 2006, the once lush, lowland rainforest had been reduced to shrubs, with only isolated fragments of old-growth forest remaining. 'Even though the area was heavily degraded, there are still somewhere between one hundred and seventy and three hundred orangutans left,' said WWF-Malaysia's Ferdinand 'Fordy' Lobinsiu. It is now a protected area and 300,000 trees have been planted to restore the ecosystem.

I felt privileged to be there in search of these magnificent orangutans. But this was tempered by an inescapable sense of loss and sadness. This is the home of some of the most iconic but endangered species on the planet. It is an area that should be protected at all costs; but for much of Borneo, the damage has already been done.

Like all these things, it is extremely complicated. It is easy to blame the Malaysian and Indonesian governments, or even to shift the blame onto local communities, but this overlooks the systemic inequalities and power imbalances that underpin the palm oil industry. The reality is that the West's insatiable appetite for cheap vegetable oil has fuelled the exponential growth of the palm oil industry. Local communities need to be put at the heart of all efforts to protect what is left of this ancient rainforest and restore what has already been chopped down. And it is not only the international demand for palm oil that is destroying our forests, but another commodity desired around the world: meat.

In 2019, I was visiting the Desana community in Brazil. To get there, I first had to fly from Brazil's capital, Brasilia, to Manaus, the largest city in the Amazon. Looking out of the window halfway through the flight, I saw fragmented landscapes that looked like a mosaic of farmland and remnants of forest. This 'arc of deforestation', extending along the eastern and southern perimeters of the Amazon, reaches to the horizon. I felt as though I was flying over it for hours.

Massive soy plantations loom large, serving as hubs for producing feed for livestock such as pigs and chickens. Meanwhile, the surrounding lands have morphed into rugged pastures for cattle grazing. I had come here to learn more about these indigenous people's spiritual connection to trees and nature, but I found myself with a shocking first view of the extent to which their incredible way of life is under threat.

What cannot be seen from the plane window is that once again this transformation is fuelled primarily by global consumer demand, originating from far-away places like the US, China, Europe and the UK. This drives the clearing of millions of acres of tropical forests each year to meet the world's growing hunger for affordable meat. At the time of writing in 2024, an area the size of Italy was burning in the Amazon – fires started intentionally by farmers to clear yet more land for grazing and crops.

The primary driver of global deforestation is our dietary habits, but what does this mean? How can a hamburger transform the future of our life on this planet? Over 16 million trees are lost every single day due to agriculture, making up at least a third of global deforestation. Beef production, in particular, is the leading cause of global forest destruction, accounting for approximately 41 per cent of land conversion. In many areas, cows outnumber humans by four times, despite cattle being non-native to Brazil.

It is not only the vast areas cleared for rearing cattle that cause deforestation, but the land needed to grow feed for them. Soy is the culprit. A small portion, about 6 per cent, is used for plant-based foods like tofu, but the majority, around 81 per cent, is grown as animal feed for chickens, pigs and dairy cows. Considering that livestock now makes up 60 per cent of all mammals on Earth, that is a lot of feed. Eating beef, in developed countries where we have easily accessible alternative options, is highly inefficient and illogical.

Cows require over 7kg of vegetation to produce 500g of flesh. On top of that it takes almost 1,000 litres of water to produce 500g of meat, in stark contrast to the mere 100 litres needed for a 500g of wheat. The production of a single hamburger consumes enough fossil fuels to drive a small car for miles. We can significantly boost the availability of food worldwide by cultivating crops for direct human consumption, instead of using them to feed livestock.

If we stop eating meat and dairy, we could cut global farmland use by more than 75 per cent.[10] That is like saving land equivalent to the United States, China, European Union and Australia combined, while still being able to feed everyone. Even if people ate just a little bit less meat, it would make a substantial difference.

By 2050, we will need to feed 9 billion people. If we stick to current food production methods, we will keep harming biodiversity and natural ecosystems. Climate change worsens competition for land, making it harder to produce enough food. Trees could be our strongest allies in the climate crisis, but only if we protect them now. The 2016 *State of Nature* report, using the newly developed Biodiversity Intactness Index, revealed a sobering reality: the UK has suffered far greater biodiversity loss over time than the global average. Ranked 29th from the bottom out of 218 countries, the UK is now one of the most nature-depleted nations in the world.

We are on the brink of a new epoch in the life of the planet.[11] Two degrees of warming is coming whether we like it or not, maybe more. By the end of this century, we will see a wave of extinctions, the tundra and Arctic ice will all but disappear, the oceans will rise and coastal cities will flood. The last generation to know a stable climate with familiar cycles has already been born. It is a bitter pill to swallow.

King Charles III said, 'Forests are the world's air-conditioning system – the lungs of the planet – and we are on the verge of switching it off.'

Deforestation accounts for about 10–15 per cent of greenhouse gas emissions globally. That is because forests are incredible storehouses of carbon. When trees are removed, the earth has less capacity to hold carbon and therefore, it is released into the atmosphere. The Amazon forest alone stores approximately 123 billion tons of carbon both above and below the surface. Unfortunately, deforestation creates a direct disequilibrium in the carbon cycle. Between 2015 and 2017, a study discovered that tree loss led to an annual gross emission of 4.8 billion tons of carbon. In other words, the amount of carbon emissions from forest loss in a single year is comparable to the emissions of 85 million cars over their entire lifespan.

If and when global temperatures rise, the consequences could be dire. An increase of 2°C could lead to disappearing glaciers, rising sea levels and extreme heat waves, causing widespread displacement and ecosystem collapse. With a rise of 3–4°C , coastal cities could vanish, agricultural failures could occur and intense weather events would become more frequent. At 6°C, rainforests and temperate areas could turn into deserts, leading to mass migration and conflict on an unimaginable scale. The ecological catastrophe could result in the sixth mass extinction event in our planet's history.

But in some places, more trees are not necessarily the answer. In 2023, I travelled to Greenland. I was filming a documentary about polar bears and the impact of a changing climate in the Arctic. This region is warming four times faster than the rest of the planet due to a process known as Arctic amplification. This is where various feedback mechanisms, such as melting ice and snow, lead to increased absorption of sunlight due to darker surfaces, increasing the speed of warming. We were staying with a remote community on the east coast, in a town called Tiniteqilaaq.

The Inuit people are the ultimate hunters, living the same traditional lifestyles that go back centuries; but with an increasingly unpredictable climate, everything is changing for them. Greenland is the largest island on Earth. Almost 80 per cent of the area is covered by ice. In its history, it has been through dramatic changes, as we learned earlier – the Vikings cut down all of its forests 700 years ago – but now, things are changing faster than ever. Computer models predict that within the next one hundred years, all sorts of new tree and shrub species could thrive in Greenland's warming climate.

At the moment only five species exist naturally, but experiments show that other species could establish themselves if given the chance. 'Greenland has . . . the potential to become a lot greener,' said lead scientist Professor Jens-Christian Svenning, from Aarhus University in Denmark. 'Forests like the coastal coniferous forests in today's Alaska and western Canada will be able to thrive in fairly large parts of Greenland, for example, with trees like Sitka spruce and lodgepole pine.'

Perhaps trees will return to Greenland despite mankind's best efforts. But is that necessarily a good thing? Polar bear populations are dwindling as it is, and may even disappear entirely by 2100 if their habitats are lost.

Greenland's ice sheet, spanning 656,000 square miles, could raise sea levels by 20 feet if melted entirely. While such a scenario is not expected soon, it underscores the threat of human-caused climate change.

Across the Arctic, this major threat comes with melting ice. The active layer of permafrost – ground that remains frozen year-round – thaws in summer and refreezes in winter, and holds about twice as much carbon as the atmosphere. Researchers now believe that for every rise of 1°C in Earth's average temperature, permafrost may release double to triple the amount of emissions from coal, oil and natural gas compared to previous estimates.

Ironically, as temperatures rise, trees spread further north, creating a problem for the Arctic ecosystem as its surface gets darker. In the northern forests, hotter temperatures lead to more intense wildfires, releasing stored carbon and methane and speeding up climate change.

Forest fires are spreading more widely and burning nearly twice as much forest as they did 20 years ago. Humans are often to blame, with the US National Parks Service estimating that around 85 per cent of its country's wildfires are a result of 'campfires left unattended, the burning of debris, equipment use and malfunctions, negligently discarded cigarettes and intentional acts of arson.'[12]

These fires have contributed to over one-quarter of all tree loss in the past two decades. In the western United States and Australia, these fires have been especially deadly and destructive. Climate change is fuelling more intense wildfires by creating hotter temperatures and drier landscapes. Extreme heatwaves are now five times more common than 150 years ago. This cycle of fires releases emissions that worsen climate change – leading to more fires and forming a dangerous feedback loop. The intense wildfires devastate crucial ecosystems, endanger properties and livelihoods, disrupt economies and communities, and emit millions of additional tons of carbon.

As trees vanish, the soil is left barren. Without the nutrients from vegetation and fallen leaves, it does not take long for the earth to turn to dust. What comes next can be catastrophic.

Nepal will always have a very special place in my heart. I have travelled there many times over the last 20 years, photographing and documenting communities and the challenges they face. Farmers in remote Nepal are dealing with climate change first-hand. They urgently need support to cope with new challenges such as insects ruining crops at higher elevations, along with landslides, droughts and the ever-present risk of floods from melting glaciers.

I saw this personally when I visited the Glacier Trust team in Nepal. A landslide had hit a small village, burying homes beneath the debris, and leaving families homeless and traumatized. As we offered what help we could, the impact of reckless logging practices was made starkly clear. I was told that the hillsides had once been lushly forested, but over time, these forests were stripped away, leaving the slopes bare. Without the trees to stabilize the soil, the area became increasingly prone to erosion and, ultimately, to mountainside instability.

Every monsoon brought with it the risk of devastating landslides that sent torrents of mud and rock crashing down onto the homes of the people living in the mountain's shadows; turning what had once been a paradisiac village into a truly perilous zone. It is a story that is being played out all across the world. The number of landslides is increasing globally, and will continue to do so until we realize and respect the role that trees play. By failing to do so, we are simply condemning more people into losing their homes, and worse, their lives.

While efforts to fight this destruction are underway, the task is monumental. Now more than ever we must safeguard our forests not just for the needs of humanity, but for all life on Earth. It is about time we reminded ourselves – there is no planet B. We have to work together with nature.

A fundamental law of biology is interconnection. Be it bacteria, fungi, insect, mammal or plant, all life relies upon other species to survive and thrive. Whether we acknowledge it or not, we are a part of this interconnection. With the rise of technology, we are more interconnected than ever before – at least, on paper. In reality, we may never have been more lonely.

Rates of loneliness have doubled since the 1980s in the West. Unsurprisingly, there was also a spike during the COVID-19 pandemic, when people were forced to isolate themselves in their homes. In an age where many people live alone, this had a tremendous impact on mental health and wellbeing, with charts clearly showing a surge in mental health problems in 2020, when lockdowns began in the UK. And it was not something that people simply recovered from the moment that lockdowns ended. The lockdown spike has become the new baseline for mental health struggles in Britain. Levels of 'happiness' have also plummeted in British youths, from a rating of 70 in 2009 to 56 in 2020.[13]

We are disconnected not only from each other but also from what we call 'the natural world' – a term that itself suggests we are separate from it. One study even concluded that children could name more Pokémon than they could wildlife species.[14]

'We are talking only to ourselves,' said theologian and ecologist Thomas Berry, in a sorrowful expression of our disconnection from the wonders all around us. 'We are not talking to the rivers, we are not listening to the wind and stars. We have broken the great conversation.'

But this does not need to be our future.

Despite the long journey ahead, the spark is there. It starts with connection, with being better people. There are beacons of hope all around: our world passed the most severe period of deforestation during the 1980s, and we have improved ever since. Today, Europe boasts a 30 per cent increase in its tree population from the previous century, with the majority of them having grown naturally. Similarly, on a global scale, our planet now has more trees than it did a decade ago.

In 1799, William Blake wrote: 'The tree which moves some to tears of joy is in the eyes of others only a green thing which stands in the way. [. . .] As a man is, so he sees.'[15]

Our world stands at a crossroads – hinging on the choices we make each day, making it our task to shape the future for generations to come – the pivotal domains of conservation and reforestation are where this comes to apotheosis. These twin pillars of environmental stewardship hold the key to addressing numerous environmental crises we find ourselves in; the pressing issues of deforestation, climate change and the loss of biodiversity.

The future of our forests is not bleak: there is a time hereafter when holistic conservation practices, community-led initiatives and innovative technologies will work in harmony to protect our planet's lungs. Through the adoption of nature-based solutions, regenerative practices and advanced monitoring technologies, we can navigate the complex challenges ahead. As Suzanne Simard says so eloquently:

> By understanding their sentient qualities, our empathy and love for trees, plants, and forests will naturally deepen and find innovative solutions. Turning to the intelligence of nature itself is the key. It's up to each and every one of us. Connect with plants you can call your own.[16]

We have explored the transition of much of humanity from forest dwellers living in harmony with nature to greedy capitalists hellbent on its destruction. But thankfully amid all the devastation, there is a glimmer of hope. And it comes from within humanity. In Zen Buddhism there is a saying that the obstacle is the path, and in the struggle lies the answer. We already possess the key to our survival, and it lies in our ability to recognize where things have gone wrong, accept it and now change what we can, in order to reconnect with the natural world.

Planting the seeds of hope will allow the rise of a new Garden of Eden – one that we can all be a part of – and it begins here, right now, leading us to a more connected future in harmony with the natural world.

Act 3

Return to Eden

Chapter 9

Tree Huggers

———

The one who plants trees knowing that he or she will never sit in their shade, has at least started to understand the meaning of life.

– attributed to Rabindranath Tagore

Beneath the glistening light that streaked through the canopy, John Evelyn basked under a sprawling oak tree that took pride of place in his personally cultivated woodland on the banks of the River Thames. It was the dawn of the 17th century, and this grove would unwittingly become the genesis of the modern environmental movement.

Inspired by his overseas travels, in particular the manicured trees that framed the houses of Rotterdam in Holland – where he was 'ravished . . . by the delicious shades and walks of stately trees' – John Evelyn decided to create his own paradise in London, a haven of ancient oaks, shrubs and wildflowers.

By this time, England was at the dawn of a new age in its history: it needed resources to fuel its exploration and power its empire. The Royal Navy, the backbone of England's maritime power, had a penchant for oak and the old groves were being cut down at a rapid rate.

Beginning to foresee the disastrous future of Britain's woodlands, Evelyn knew he had to do something about it. His debut book, *Sylva*, came from a request by the Commissioners of the Navy to the Royal Society for a report on forestry. Through *Sylva*, Evelyn inspired landowners to plant new trees, both at home, on city streets and in formal gardens. His book did not merely provide practical advice, but also interweaved literary stories and poems from classical texts, thus creating a poignant sonnet to the forests of Britain.

If you take a walk in the woodlands and forests of the British countryside today, it is more than likely the care and management strategies that maintain the flourishing vegetation stemmed directly from the methods of John Evelyn. Ever since, his work has been a fundamental reference point for tree lovers. The first step to reconnecting with nature is to spend time in it, in a bid to nurture a new understanding of our role as stewards and guardians. As Evelyn understood with his work in the British woodlands, the future lies in what we preserve today.

But the seeds of conservation have grown far beyond Britain's borders. Protecting our natural world for future generations is not just a local endeavour – it is a global responsibility. Seeds literally provide hope for the next generation.

In this spirit, deep inside a mountain on a remote Norwegian island, the Svalbard Global Seed Vault houses millions of seeds from around the globe. Born as a collaboration between the Norwegian government, the Crop Trust and the Nordic Genetic Resource Center, this 'doomsday' vault acts as a fail-safe for preserving agricultural biodiversity on our planet, and is designed to withstand nuclear

Armageddon. The precious seeds are stored at a constant temperature of -18°C (-0.4°F), ensuring that they remain viable for decades, and sometimes even longer.

The purpose of the vault is to safeguard the genetic diversity of crops, which is essential for breeding new varieties that are resistant to diseases, pests and climate change. It is a symbol of global cooperation, and a testament to the importance of preserving our agricultural heritage. It also serves as a vital resource for researchers, breeders and farmers worldwide, ensuring the survival of our food crops for future generations.

But preserving seeds from different locations around the world is not a new concept. While the doomsday vault sounds like something that a band of nomadic survivors would seek out in a science-fiction movie, the Chelsea Physic Garden feels far more accessible. Just nine years after the publication of *Sylva*, London's oldest botanical garden was set up. Nestled between Chelsea Embankment and the Royal Hospital, the garden dates back to 1673. It was here that the first Lebanese cedars were planted in England.

Showing me around, Nell Jones, head of plant collections, points to some of the 4,500 species of plants and trees. 'This place has been a teaching garden for over 350 years. It was set up by the apothecaries to understand more about useful plants.'

Remember the tea plants and the role that tea played in the British Empire? Well, we can thank (or blame) Robert Fortune for his role in that chapter. Fortune was one of the most famous British plant hunters. He was the curator of the Chelsea Physic Garden from 1846 to 1848. He left the post when he was recruited by the British East India Company to play his part in horticultural espionage. Fortune was sent to China to obtain the tea plant

Camellia sinensis illegally, when tea production was a closely guarded secret.

Nell told me, 'Did you know that the study of plants was mandatory in the field of medicine until 1895?' She chuckled. 'I think that's about when things started going wrong.'

England is poorly endowed with native trees when compared to mainland Europe, and the comparison is even greater with the tropics and botanical hotspots such as Japan. Only around 30 tree species managed to recolonize our shores in the wake of the receding ice sheets after the last ice age. Our separation on a small island slammed the door shut to latecomers, and so the present rich diversity can be fairly accredited to the British people and their fondness for planting trees. Centuries of human collection, seeding and cultivation have resulted in a wide and beautiful arboreal heritage that lasts to this day.

Even in Roman times, travellers brought with them seeds and saplings from Italy to remind them of home – walnut and sweet chestnut now abound. Successive waves of immigrants, traders and herbalists added to the variety of stock that we now take for granted.

A rising interest in exotic plants led to a new phenomenon, as public botanical gardens sprung up all over the country, particularly from the late 17th century to the early 18th century. Monasteries, convents and royal houses had long kept private gardens, walled enclosures, or even large plantations in which plants were cultivated for food or their medicinal value. But these new botanical gardens featured a vast array of flora from around the world. Exotic trees provided not only new nutrition but new medicine, and scores of 'physic gardens' supplied drugs to apothecaries.

Up and down the land, country homes vied for the best gardens as aristocrats styled their own arboretums. Trees became collectors' items, especially those with political or religious significance. The Lebanese cedar became a sought-after prize to show one's Christian credentials, just as palms were uprooted from the Holy Land and planted in London to instil some godly fortune, and rhododendrons from India served to brighten up many an English garden.

Saving seeds is an inconspicuous, heroic act for safeguarding the biological integrity of the planet. In Victorian England, this practice took the fashion of transporting seeds from California's iconic redwoods to be planted in gardens and estates across the UK. The impact was little acknowledged at the time – redwoods do not start off as giants after all – but now less than 150 years later, at around 40 metres (131 feet) their crowns are showing over the typical English treeline, leading botanists to take renewed interest in the place of redwoods in the UK. One could argue that these trendy Victorians were some of the first unwitting environmentalists.

Currently, these giants are still juvenile. If they grow into their true size, they can reach three times their current height and live for thousands of years or more. Giant sequoias (*Sequoiadendron giganteum*) at their pinnacle can reach around 250 feet and live over 3,000 years.

The experience of standing under a grove of redwoods is truly mind-altering. From my experience, these trees have the power to reorient our view of nature. They humble us and evoke a sense of awe and wonder at their sheer size and the untold stories within their canopy. Writer John Steinbeck was also awestruck by these majestic trees, calling them 'ambassadors from another time'.

When you stand at the base of a mature redwood, the top of the tree is obscured by an understory of vegetation hundreds of feet above the forest floor, with ferns, lichen and a whole micro-ecosystem of niche species all the way up the trunk. Even saplings and bushes of plants like currants, huckleberries and hemlock have been found living high in redwood branches. In one study, one hundred different species were found living on one single tree! It is as if the redwoods are single-handedly conspiring to keep some of nature out of human hands to safeguard species on their own terms.

By an interesting twist of fate, the proclivity of the Victorian era for redwood seed exchange has resulted in the UK being the home to more than half a million of these giants today. Those numbers overshadow the dwindling amount of redwoods in California, where the 80,000 survivors are struggling with the compounding impacts of prolonged drought, climate change, bark beetle invasions, a legacy of counterproductive forestry practices and increasingly intensive fires that ravage the region now every summer.

In the mid-19th century, redwood seeds were first brought to the UK around the same time that Charles Lutwidge Dodgson – better known by his pen name, Lewis Carroll – was walking through the Oxford Botanical Gardens with Alice Liddell, the young girl who inspired *Alice's Adventures in Wonderland* (1865). Carroll, a mathematics professor at Oxford, often visited the gardens with Alice and her sisters, sparking the whimsical, imaginative spirit found in his famous novel. Today, much like Alice, we can share in this sense of wonder by shrinking down (in spirit) and walking beneath the towering redwoods that now grace the Oxford Botanical Gardens, Kew Gardens, and many other places across the UK.

'Kew Gardens are historic,' said Richard Deverell, the director of Kew, showing me around the verdant gardens. 'We have about 20,000 species here in 300 acres making it possibly the most biodiverse spot on the entire planet. It's a global collection, with plants from all over the world, from ones you're familiar with like the English oak, through to some exquisitely rare plants, some of which are extinct in the wild.'

He pointed to one particular oak tree. 'I love the fact that this individual tree has witnessed everyone who has ever walked down this vista since it was created in the 1840s. Kings, Queens, Prime Ministers, this tree has seen them all.'

Kew, he told me, is a 'repository of global knowledge which can help solve problems like finding climate resilient trees, plants and crops – nature-based solutions which can help feed the world.'

During the Industrial Revolution – and perhaps because of it – many new tree species ended up flourishing in Britain. Nowadays you can visit hundreds of these wonderful collections, many of which are taken care of by the National Trust (NT). 'The UK has got more ancient trees than any other country in Europe,' said John Deakin, the NT's Head of Trees and Woodlands. 'We have a huge responsibility to safeguard that kind of biodiversity and nature that depends on those trees, and also the cultural significance of those trees and the connections with people they've had for many years.'

He is absolutely right. And preservation starts in our back gardens; they nurture us with the fragrance of flowers and a home for the bees. Gardens provide a sanctuary for pleasure and contemplation. In many languages the word for garden signifies an 'enclosure', evoking the idea of a personal place full of mystery and secrets known only to the gardener. The garden paradise is the imagined locus of our beginning and end, a sacred place of wonders, a world full of wholesome potential and divine creation. It is no surprise then that Muslims speak of gardens as states of bliss and call Allah 'The Gardener'.

As an Englishman, I am occasionally forced to admit that the French have pioneered a great many lovely things. Nevertheless, I am not a fan of their baroque gardens. Yes, they are majestic to look at, but they are too orderly and staid. They also require *constant* maintenance. Each hedge must be trimmed, each bush must be clipped, and the lawns must be regimentally maintained. French gardens embody a strong desire to order and control nature, reflecting humanity's supposed dominion over the natural world.

They are all about carving up nature to fit human ideals of beauty and order, rather than adjusting man and his ways to the rhythm of the natural world. Unsurprisingly, this style of garden came out of the Age of Reason, an intellectual and cultural movement that spanned the 17th and 18th centuries in Europe, which emphasized reason, logic and science.

The English are a more practical people, who tend to prefer a certain level of anarchy in their gardens. My favourites are those designed to work alongside nature's rhythms, as opposed to against them. The idea is not to conquer the natural state of things, but simply to guide it along its way. We like to seed our gardens with a dizzying array of plant life that attracts an even broader array of wildlife. The process is spontaneous and liberating, and often at odds with the French way of turning bushes into cones and lollies.

This contrast in gardening techniques mirrors a broader philosophical divergence in how various cultures approach their relationship with the natural world. A walk through a traditional English garden is an exploration – of hidden pathways and curious discoveries around every corner. The worldview that we carry with us impacts our day-to-day relationship with the natural world. Are we here as partners, friends and co-creators, or as dominators?

The idea of caring for the environment and protecting the trees might have once been novel for Europeans, but it was not elsewhere.

India is home to a remarkable community of people known as the Bishnois. For centuries, the Bishnois have forged a unique and deeply reverent relationship with trees. The Bishnois originally came together during the 15th century under the leadership of Guru Jambheshwar, a visionary spiritual leader. He laid down 29 principles to help guide the lives of his followers, emphasizing compassion, non-violence and great esteem for all life. These principles, collectively known as the Bishnoi Dharma, would become the cornerstone of the community's environmental ethics.

At the heart of the Bishnoi Dharma is the concept of 'Sacred Trees' or 'Khejri' trees, specifically the *Prosopis cineraria* – a drought-resistant, desert tree native to the region. These trees are pillars in the Bishnoi way of life, and their protection is an integral duty of the community.

One of the most iconic and deeply moving stories of the Bishnois' relationship with trees revolves around the legendary sacrifice of a young woman by the name of Amrita Devi. In 1730, a party of royal officers, sent by the Maharaja of Jodhpur, arrived in the village of Khejarli with the intent to fell Khejri trees for the construction of a palace. In response, Amrita Devi led a fierce protest, wrapping herself around the tree to stop the foresters from cutting it down. The tale goes that as she hugged the tree, foresters attacked both Amrita and the trees with their axes. She refused to compromise, stating, 'If a tree is saved at the cost of one's head, it's worth it.'

Another 362 Bishnois followed suit and wrapped themselves around the trees. Tragically, all of them, including Amrita Devi and her daughters, were killed in their steadfast resistance. Only at this degree of bloodshed did the foresters relent. This became a turning point in the commitment of the Bishnois to safeguarding their right to protect their environment, and marked the beginning of India's environmental conservation movement, long before the modern concept of conservation took root.

A temple was later erected in tribute to those Bishnois who gave their lives for the Khejri trees. The sacrifice of Amrita Devi and her companions sent shockwaves through the region and as news of their bravery spread, rulers of the time recognized the sanctity of the Bishnois' beliefs and enacted laws to protect the environments that held Bishnoi communities. Today many Bishnois make the pilgrimage every September in honour of this history and continue to live by the principles laid down all those years ago by Guru Jambheshwar.

These days the Bishnois' commitment to tree protection extends beyond their local communities, as they actively patrol and guard the regions surrounding their villages to prevent illegal logging and tree felling. Their vigilance has resulted in the apprehension of poachers and timber smugglers, helping to maintain the fragile ecological balance of the Thar Desert. In their villages, it is common to find groves of sacred trees where peacocks, deer and other wildlife gather freely. As a result, these groves have become oases of biodiversity in an otherwise harsh desert landscape.

The Bishnois' environmental ethics encompass all aspects of nature, not just the trees – bestowing respect upon all living beings, from insects to goats, as well as holding a special place for the elements. Their philosophy of non-violence extends to their daily lives, influencing their diet and the way they interact with the world around them.

India has played a long-standing role in inspiring environmentalism. Throughout the 1970s, in the region of Uttar Pradesh in the Himalayas, a movement known as the Chipko (Hindi: चिपको आन्दोलन translating to 'hugging movement') took root in response to deforestation.

When rumours stirred that the government planned to fell sacred trees in her local village of Reni, Gaura Devi began rallying forces to face these threats. She was joined by a group of 27 other women as they huddled together, arms embracing the trees as family for four days, before the loggers retreated. The effects of their vigour rippled across the region, resulting in the forest becoming classified an ecologically sensitive area, and they retained a ten-year ban on all tree-felling in the area.

The essence of the Chipko movement unfurled across India, with tree-hugging resistances taking root in villages across the land; elevating to higher levels after Mrs Indira Gandhi, India's leader at the time, proclaimed the Chipko movement as a work of 'moral conscience'.

Often it is the actions of a small number of determined individuals who can inspire social and moral improvement on huge scales. We are all stewards of nature.

From the Scottish missionary David Livingstone, who became a vocal protector of the baobab trees around Victoria Falls,[1] to the philosopher and naturalist John Muir, who became a wilderness advocate in the USA, travellers and explorers have often served as creative kindling to pioneers of other kinds. Take Alexander von Humboldt, who traversed the Orinoco, one of the longest rivers in South America, in the late 18th century. Alexander had no other goals than to make astronomical observations, measure the mountains, and carry out any research tasks that he deemed as scientifically valuable.[2]

His exploits sparked the interest of Charles Darwin, who may never have made his way onto the HMS *Beagle* without exposure to von Humboldt's findings. The curiosity of these Western pioneers helped to create anticipation back home and popularised natural history and taxonomy to the masses in the West. This laid the groundwork for the new wave of naturalism in the Victorian era, when men and women would bring back intriguing collections of plants and seeds from distant lands, ushering in a trend of rearing exotic trees in British homes and gardens.

Even at the peak arrogance of the Industrial Revolution, not everyone was thrilled about the excessive consumption of fuel and the demolition of native forests. In 1826, an anonymous book started appearing on bookshelves. Ominously named *An Essay on the Principle of Population*, it detailed a mathematical principle that predicted the numerous food security crises the world currently faces. It explained that unless birth rates drastically declined, food supplies would never align with that of the exponential birth rates that began after the Industrial Revolution. The book spoke of the threats of an eternally growing population, such as poverty, the result of a growing supply of labour but inevitably dwindling wages.

This is now known as the Malthusian principle – the anonymous author being Thomas Malthus. It is said that the publication of this book served as an inspiration for Charles Darwin's and Alfred Russel Wallace's theory of natural selection. It was starting to click into place that humanity was running rampant with little regard for non-human life. Jeremy Bentham was the first to advocate for limitations on how one life can impact another. His moral theory, which asserts that 'actions are right insofar as they produce pleasure or prevent pain', laid the foundation for his manifesto aimed at advancing the rights of non-human animals[3] and is the first groundwork for modern animal rights movements.

It was around this time that the first awareness of global warming began. Jean Baptiste Fourier, a French mathematician, sought to answer why the planet does not keep heating up as it receives sunlight – and how come it does not overheat.[4] His work was instrumental in identifying carbon dioxide and methane as heat-trapping gases in the atmosphere.

This period was marked by significant social and spiritual change. As people increasingly questioned Christianity and Catholicism, new beliefs began to emerge. In 1836, there came an additional wave in the shifting consciousness of Victorian England, by the hand of Ralph Waldo Emerson. In his essay, 'Nature', Emerson explored why people often fail to appreciate nature's beauty. He did this through a division of nature into four distinct purposes through the human lens: commodity, beauty, language and discipline. These purposes aimed to define the different modalities in which humans use nature for their basic resources, their desire for joy, communication, and their understanding of the external world.

Emerson also coined the belief of transcendentalism within nature:[5] that people and all of nature are inherently good, while our external environment in the form of manufactured societies has corrupted our innate wholesome spirits. At heart, this was a call for an elevation of the importance of the natural world in modern society, with a proposed limit on human industrial expansion into wilderness areas. Nearly 200 years later, we still seem to struggle with the concepts he posed.

Some 20 years later, a man by the name of Henry David Thoreau would produce his magnum opus, *Walden*, or *Life in the Woods*, in which he detailed the period he spent living in a self-constructed cabin on the land of his dear friend, Emerson. He wrote of the influence of solitude on connection to the natural world and contrasted his freedom to the imprisonment of others, who devote their lives to striving for material affluence.

After spending a night in jail for refusing to pay poll taxes, he wrote 'Civil Disobedience'. This essay is often considered a foundational text in the modern tradition of environmental and social protest.

A few decades later, George Perkins Marsh published *Man and Nature*.[6] In this groundbreaking work, Marsh criticized humanity's detrimental impact on the natural world and condemned our activities as harmful to the very environment that sustains life. He reiterated the fact that if we do not find a way to restore and sustain these resources we have pillaged, we will see our own demise. Marsh challenged the prospect of an inexhaustible Earth, and drew from lessons in history, such as the societal collapse of ancient Mediterranean civilizations, to speak on the harsh reality of the dangers of ruining the health of our soils in intensive agriculture. *Man and Nature* is famed as one of the starting guns of the modern conservation movement.

In the mid-19th century, among the forested Sierra Nevada Mountains, one man would embark upon a journey that changed the face of conservation across the United States. John Muir, a Scottish-born American naturalist, set foot in the region of Yosemite for the first time in 1868, having heard whispers of its mythical grandeur. Entering the valley, surrounded by imposing cliffs and plunging waterfalls, he felt shockwaves of awe and gazed at the monumental granite towers that stretched up to the sky. He marvelled at the unimagined plant life, surging rivers, and the light as it danced through vast meadows. Yet above all, it was the trees that captured his heart.

One evening, as the last of the sun's rays tucked below the horizon, bathing the valley in a golden glow, Muir wandered into a grove of giant sequoias. Feeling a profound connection, he placed his hand on one of the ancient trees, tracing the rough, weathered bark. Inspired by this moment, Muir wrote his essays 'My First Summer in the Sierra' and 'The Mountains of California', which offer insights into his transformative experiences as an explorer.

'It seemed the very central heart of the world, sending pulses of harmony and response through every fibre of me,' Muir later wrote. 'The colossal specimen seemed to be telling, in mighty tones, the story of the Earth.'

At this time, the notion of a National Park had little meaning. It merely acted as a label for an 'area of interest'. Yet, Muir's writings were the catalyst of change. His descriptions and iterations of Yellowstone and the Sierra Nevada began to do something to the minds of the American people, allowing them to see these wildernesses as less of an objective interest, and more something that inherently speaks to the values at the core of the United States. Muir was transforming Yellowstone into a poignant symbol of nature as a vast, holistic system, embodying the divine and representing the so-called American ideals:

Harmony rules supreme. *Linnaea* [twinflower] hangs her twin bells over the rugged edges of the cliffs, forests, and gardens are spread in lavish beauty roundabout, the nuts and berries ripen well, making good pastures for the birds and bees, and the bears also, and elk, and deer, and buffalo – God's cattle – all find food and are at home in the strange wilderness and make part and parcel of the whole.

Muir was the founding president of the Sierra Club,[7] a group with the original sole intention of sponsoring outdoor expeditions in the mountain regions of the Pacific Coast. But Muir quickly began involving the club in political action to further nature conservation.

It was during Muir's time spent in Yellowstone that President Theodore Roosevelt paid him a visit.[8] The pair undertook a three-day expedition where they camped beneath the Mariposa Grove of giant sequoias and endured a snowstorm with five feet of snow.

Muir ignited a deeper appreciation for the beauty and magnificence of Yosemite in Roosevelt, which led him to expand federal protection of Yosemite, and inspired him to sign into existence 5 further national parks, 18 national monuments, 55 national bird sanctuaries and wildlife refuges, and 150 national forests.

Roosevelt's presidency would go on to change the course of conservation in the US, with the establishment in 1905 of the United States Forest Service. The Wilderness Act of 1964 would later follow, which called into action the safeguarding of over 9 million acres of wilderness areas, leaving an indelible mark on the conservation movement, not just in the US but globally. Roosevelt's positive legacy on the environment can be directly attributed to the fact that he spent time in the forests himself. He saw with his own eyes the value of wild spaces and found a personal connection to the natural world.

The 1960s arrived at the terminus of a period of discrimination and restriction, bringing with it a new dawn of counterculture;

summers of love, feminism, and perhaps most paramount, the dawning of the modern environmentalist movement. Among the cultural shifts of this time was a rejection of mass consumerism and the 'rat race' of modern living, as young adults, disillusioned by post-war urbanization and becoming increasingly entrenched in society, began to seek solace in nature once more. Looking to the naturalist icons of the previous century, such as Henry David Thoreau, a whole generation sought to reconnect with a slower pace of life.

This movement became synonymous with care-free living and the celebration of the arts and thus, the hippy movement was born. Taking inspiration from the ashrams of India, communes began to pop up in rural areas, establishing an emphasis on self-sustaining, off-grid living with small-holding agriculture that opposed the industrial farming of the war.

These lifestyles continued to blossom and gained mainstream popularity with the release of Stewart Brand's *Whole Earth Catalog* in 1968, which rapidly became a bible for the back-to-land movement, providing easy-to-follow guides on tools, resources and essential techniques for thriving in a self-sustaining home. It played a crucial role in popularizing environmental responsibility among individuals and influencing the development of eco-conscious lifestyles.

The first Woodstock Music and Art Fair[9] of 1969 became a prominent symbol of the counterculture of the time, as benevolent hordes gathered in serene spirit on a dairy farm in upstate New York. Just a year later, the inaugural Earth Day was celebrated on 22 April 1970. Millions took part, set against the backdrop of a recent oil spill in Santa Barbara, and the first images of Earth from space – the 'Blue Marble' photographed from Apollo 17 – were beamed around the globe, affirming the notion of our planet's fragility. Earth Day's success was one of the major factors that resulted in the establishment of the Environmental Protection Agency (EPA) and the passage of key environmental legislation, such as the Clean Air Act and the Clean Water Act.

Emerging from the societal upheaval of the 1960s and early 1970s, a new generation of environmentally conscious activists began to spread their message as access to information improved and media platforms became more global.

The 1980s saw the creation of the EPA's Superfund program and strengthened protections under the Endangered Species Act. By the 1990s, activism expanded to include global issues like climate change, culminating in the Earth Summit of 1992, which marked a pivotal moment in international environmental policy.

In the winter of 1997 amid the coastal redwoods of California, a young woman named Julia Hill captured the world's attention and inspired a new wave of global activism. The 23-year-old from Arkansas found herself at the centre of a political and environmental storm when she chose to take a stand against the impending annihilation of a pristine redwood forest by the logging company Pacific Lumber.

With a small group of like-minded activists, Julia climbed a 1,000-year-old redwood, which had been named 'Luna' after the rising moon. High above the forest floor, she established a home in the canopy, determined to protect Luna and the surrounding trees from the clutches of the loggers. Julia intended to remain in the tree for three weeks – a fearless and symbolic act of defiance against the logging company. However, those three weeks turned into two years. Julia and Luna became the figureheads of a remarkable environmental confrontation.

Living among the treetop canopy, Julia endured brutal winds and storms, intense heat and exhaustion, isolation and a multitude of emotional challenges. She relied on the kindness of supporters bringing her water, food and supplies via a rudimentary rope-and-pulley system. Through it all, she remained steadfast in her mission to protect Luna and the redwood forest.

Julia's tree-sit captured the minds of the world media. News outlets, environmentalists and curious onlookers flocked to the forest to witness her remarkable act of protest. The media dubbed her 'Butterfly', due to her deep, delicate and endearing connection to the creatures that frequented the tree. Julia used her newfound platform to raise awareness about the importance of preserving ancient forests and the devastating consequences of deforestation.

The battle for Luna and the surrounding forest became an emblematic struggle between the forces of industry and the defenders of nature, with legal battles and negotiations. Her display of dedication and unwavering commitment led to the formation of an agreement that saved the tree and protected a significant portion of the forest from logging activity. The triumph of this campaign showed the world that individuals *do* have agency and a role to play by challenging the damaging choices made by corporations, inspiring countless others to take action in defence of our natural world.

Julia's remarkable journey did not end with her descent from Luna. She continues to advocate for environmental causes and has since dedicated her life to raising awareness about deforestation, climate change, and the urgent need for action to protect and preserve ancient forests.

I spoke with Julia 25 years after her heroic stance. I asked her, now that she is 50, to reflect on her actions:

After Luna was attacked, we didn't know if she'd make it until five years later – but what was beautiful is that all kinds of people came together to save Luna's life. Structure engineers, biologists, botanists, indigenous people. In the years that followed, California had some pretty severe droughts – but Luna never wavered. It's been 25 years. Every spring she has new growth. She's defying science. These people who attacked Luna were trying to attack what Luna stands for – but all they did was make her stronger.

Even after all this time, the passion in her voice remains unwavering:

Living in Luna taught me to think of ourselves as ancestors of the future. I don't want people to look at people like me as hope for our world, they need to look at themselves too. It's not about hope, or an absence of hope. I'm going to live and love in action as best I can.

Julia's story is one of immense fortitude in the face of environmental misjustice. Thankfully she was able to bring activism to the masses by utilizing the media to create more awareness to her cause.

But no one else has brought multispecies empathy to the masses more successfully than Sir David Attenborough, who has been championing natural history for over 65 years, a consistent voice for those who cannot speak.

Marked by his tenacity for exploration and deep appreciation for species and ecosystems, his life has brought the wonders of the natural world to a broad audience. He has popularized and shared with enthusiasts what many niche researchers and ecologists could only dream about. His ability to incite emotional investment, curiosity and intrigue about these otherwise esoteric subjects – everything from parasitic fungi, algae and insects by the hundreds – has garnered gratitude across the globe.

I would not be able to begin to estimate how many people he has influenced over his years in broadcasting, but one thing is for sure, walk into any Life Sciences department at a university and you will find posters, reference books and individuals that would not be there without him. His lifetime is one that has spanned the greatest one hundred years of change ever known, experiencing still pristine ecosystems and newly discovered species to unfortunately seeing their decline and loss. The existential pain that a life filled with such beauty and loss must bring is unimaginable, yet it is what galvanizes his role as the patriarch of 21st-century environmentalism.

The last 30 years have seen the establishment of countless charities, non-profit and local organizations to combat habitat loss and deforestation, which if executed in a thorough and ecologically viable manner, is one of the greatest ways to fight the issue. Environmental activists are essential to protecting trees, but in many areas of the world it is a dangerous movement in which to be involved.

Drawn to the plight of the rainforest, British journalist Dom Philips moved to Brazil back in 2007. There, his life was spent assimilating to the country and its ways, understanding the intricacies of its social ecosystem, and documenting the troubles afflicting its greatest source of conflict, and those desperately trying to save it – the Amazon.

In 2022, Dom began to write a book called *How to Save the Amazon,* warning of the threats of illegal fishing, logging and drug smuggling in the region. In June of that year, Dom and Bruno Pereira, an indigenist and former government official, embarked on a trip they had taken numerous times before, travelling by boat through the

Javari Valley of the Amazon on the Itaquai River. Pereira, accustomed to the threats of the illegal workers, always carried a gun when he was out instructing local people on how to patrol their land for these illegal activities.

They were accompanied by locals, including a man named Caboco, who was later found to be linked to an illegal fishing syndicate, and disappeared with Dom and Bruno shortly after they set off. For ten days they were missing, only to be discovered shot and buried in the muddy ground of the jungle, their bodies desecrated. Almost immediately afterwards, two brothers belonging to the same illegal fishing syndicate, local to the Itaquai, confessed to the murders.

Today, Dom and Bruno's names are raised on placards petitioning for indigenous rights, environmental and journalistic protection and constantly included in policy debates about elevating state presence in the Amazon. It is individuals like Dom and Bruno who are to be thanked for their work in publicizing the wrongdoings of humans against our forests and their communities. The pair are now immortalized as martyrs of environmental destruction, forever bringing light to those less recognized victims of conservation crimes.

Environmental activists remain at risk, with hundreds of people murdered each year in the Amazon. And yet, many more continue to risk their lives, their families, their economies and their reputations for a higher cause – that of protecting nature. And what could be more noble? As Theodore Roosevelt once said:

It is not the critic who counts: not the man who points out how the strong man stumbles or where the doer of deeds could have done better. The credit belongs to the man who is actually in the arena, whose face is marred by dust and sweat and blood . . . who spends himself in a worthy cause; who, at the best, knows, in the end, the triumph of high achievement, and who, at the worst, if he fails, at least he fails while daring greatly, so that his place shall never be with those cold and timid souls who knew neither victory nor defeat.

In the middle of the Pacific Ocean lies the Hawaiian island of Maui. On arrival, you are greeted by the smells of frangipani flowers and the bright reds and pinks of hibiscus. The Ma'a'a wind skims across the sea and graces the old port town of Lahaina, rustling the foxtail, Caryota and Areca palm trees and coconuts. The Kakui and Macadamia trees are laden with nuts ready to be picked for eating and for making oil. Or at least they used to.

In the summer of 2023, this beautiful part of Hawaii was ravaged by wildfires. More than 2,000 buildings and 100 lives were lost, 11,000 residents were displaced, and the coastal town of Lahaina was burned to ruin. The fire left in its wake block after block of debris and ash, the town's roads lined with abandoned cars that had been reduced to twisted metal frames. Across Lahaina, blackened trees stood as silent mourners of the devastated community.

I visited the beleaguered town in January 2024 and was shown around by the local arborist, Tim Griffith. Amid the carnage, one particular tree stood out. At the centre of Lahaina town is the largest banyan tree in the USA, spanning two acres, roughly the size of a city block, and rising more than 60 feet. Indian missionaries presented the mammoth tree as a gift in 1873. Ever since, it has been a beloved sight, growing rapidly to offer shade for the community and attracting tourists from all over the world.

Recently, before the fires struck, the town marked the banyan's 150th birthday with a cake. It is a tree that is very much a celebrated member of the community. Even the Hawaiian King Kamehameha III held a birthday party under the tree in 1886. 'It's a gathering place,' Tim told me. 'For generations people have been coming and enjoying this tree.'

Everything in Lahaina was destroyed, but miraculously the beloved banyan tree survived. I stood with Tim in the middle of the old tree, its 16 serpentine trunks and countless branches surrounding us. It is a sight to behold. Standing out against the desolate, burned backdrop, its beautiful new leaves and shoots swayed slightly in the breeze.

'We don't fully understand how it managed to remain alive,' Tim said, 'but for the community to see its resiliency, to see that it survived the fire and it's putting on new growth, it just gives hope to the community that the same thing will happen for us, that we will continue to be able to live here and we will continue to grow.' It was only a few months after the terrible fires, and the wounds were still fresh and raw. 'If the tree can survive, then so will we.'

It was as if the Lahaina banyan tree, looked after by the Hawaiian people for years, had reversed its role and was now offering comfort to its carers in return. After all the loss, it still stood strong, quietly giving solace to the community.

Trees evoke a unique sense of familiarity, with their identifiable individuality and often human-like forms, featuring limbs and a trunk. The very fact that you can hug a tree offers comfort. Perhaps there is a tree-hugger deep down in all of us; we simply need to have the courage to release it.

Trees are stewards of us as much as we are stewards of them. Our stewardship can come in all shapes and sizes. People like Tim Griffith, who dedicate their lives to caring for trees, are an inspiration, but being a champion for trees is something that all of us can do, even if it is on a more local level.

Not all heroes wear capes, so the saying goes. And in the writing of this book I have been astounded by how many people, friends and strangers alike, have a deep connection to trees, even if they do not shout about it.

'Don't tell anyone my real name,' said my friend. 'Just call me "The Kernel".'

It took me a second to get it. 'Oh, yes.' Colonel/Kernel.

'The Kernel of Truth', as he goes by in underground planting circles, is in fact a high-ranking officer in His Majesty's Armed Forces. We had met many years before on paratrooper selection and he had gone on to gain promotions and senior leadership roles, and now had a very serious job doing important government things.

'I hear you're writing about trees?' he whispered to me one day in the pub.

The Kernel carried on, evidently enthused at our hitherto secret shared passion.

'Good work on joining the fight. I'm settled down now, in the suburbs with my kids, but we're leading our own little tree rebellion.'

I must confess I was a little surprised. The Kernel was the last person I could imagine running around with a watering can in a pair of gardening gloves. I had no idea that he was particularly fond of nature. He then proceeded to tell me one of the most ingenious, silent acts of rebellion I have ever heard.

'So me and the kids grow saplings in the greenhouse, and then when they're big enough, we take them off to local parks and plant them in the dead of night. I even put sniper tape around them so the other guerrilla planters know what's what, and some of the yellow tape too, so the council thinks they've planted them themselves.' He chuckled. 'They must give themselves a right pat on the back.'

With a conspiratorial wink, he carried on. 'We go to Windsor Park and gather acorns and plant them anywhere we can. Roundabouts, hedges, even in central London. Over the years we've planted hundreds. We go at weekends to visit them and watch them grow up. It's become our family tradition. I guess it's my way of protest.'

'Guerrilla tree planting', or guerrilla gardening, is controversial when unauthorized, but not technically illegal. The planting of saplings in public or private spaces can certainly improve the health of the surrounding environment and increase biodiversity within urban regions. Common choices for guerrilla tree-planting activities include parks, vacant lots and properties, along the verges of streets and roads, and common green spaces that lack council supervision.

Seed bombing is another subtype of guerrilla tree-planting – the distribution of seeds, normally bundled into cotton cloth and thrown into areas where the planter wants trees to grow. Creating seed bombs containing native wildflower seeds is another, as these help to provide small oases for pollinating insects during the spring and summer months. Introducing any native flora to an urban area is good for the mental wellbeing of those who live nearby, as these small pockets of life boost the moods of those who regularly pass them.[10]

The Kernel explained that his fellow secret seeders often carry out these activities at night to avoid any repercussions, and while it would be unlikely that anyone took real offence at the activities, he acknowledged that sometimes he might be accused of trespassing.

Whether you might agree with the tactics or not, these sorts of activities can unite communities with landowners and local authorities, and seeing the benefits of this illicit planting often encourages them to develop official greening programmes and legitimize their efforts.

Now, I am not suggesting everyone becomes an undercover gardener, but the Kernel's words did seed some hope for me. That personal connection with nature can inspire change, regardless of your background or day job. Immersion in nature, whether it is getting your hands dirty, tuning into the sounds of the forest, or just looking at their magnificence for what might be the first time, will help you to shift your perspective and notice the life around you.

The Kernel is as far from being a tie-dye-wearing tree hugger as one can get, but this was his resistance against an industrialized world that pushes away the ability to connect to Mother Earth. It was a way for him to teach his children to love and revere nature, to be proud of watching their trees grow, and in a greater sense, to be champions of the environment for the rest of their lives – albeit in a rebellious way.

Of course, there are plenty of official ways that do not involve sneaking around in the dead of night. Volunteering with organizations like the National Trust or simply taking action in your own garden can be the start of something big. 'If everybody in the UK planted just one tree, that would more than double our woodland ambition, so in effect that would create something in the region of 60,000 hectares. What a contribution that would be, now imagine what would happen if people planted three, four or five,' John Deakin of the National Trust told me.

Thanks to the singular and communal efforts of people all around the world, the idea of stewardship is coming back. The industrialized nations are finally beginning to reconnect with what we lost when we left the Garden of Eden. But we must remember that some people never left – and it is from them that we might learn the real secrets to becoming true custodians of the trees.

Chapter 10

Seeds of Wisdom

———

*Look deep into nature, and then you will start to understand
everything better.*

– Albert Einstein, to Margot Einstein after his
sister Maja's death in 1951

The beach resort of Sharm El Sheik in Egypt seemed an odd place to
come and learn about trees, given that the only evidence of greenery
were planted palms that shaded tourists in hotel enclosures and lined
a few of the sandy roads. But in November 2022 I had been invited to
speak at COP27 – the global climate change conference – where
world experts, scientists and politicians gathered to discuss how
much, or how little, was being done to combat the climate emergency.
Under the sponsorship of Coca-Cola (somewhat oxymoronic given
the brand's dubious environmental record), the event seemed to do
little more than garner widespread criticism from the world's press
and boycotts from activists like Greta Thunberg.

However, I was not interested in political posturing or virtue-
signalling. I had come to meet with a very specific group of people, the
Indigenous Community Forum.

Leaders from across the forested world had come together to amplify the voices of their communities, all trying to spread a very similar message. I had the joy of meeting men and women from all parts of the globe – the Congo, the Amazon, India, New Zealand, Mexico and Guatemala – all dressed divinely in technicolour traditional clothing.

I had been hoping to see Benki Piyãko again, the leader of the Asháninka, who had set me off on this path in the first place, but he was not there this time. In his place was Isku'kua and Rasu from the neighbouring Yawanawá tribe, as well as elders from the Huni Kuin. There were representatives from North American tribes, First Nations people of Canada and Queen Diambi from the Congolese Bakwa Luntu. I attended lectures, panels and discussion groups in the incongruous setting of hotel conference rooms.

The final evening saw a fire circle, where prayers were shared, stories were told and the graces given to ancestors. It was a profoundly beautiful experience, listening as the flames licked at the coals, casting flickering light on the faces of these individuals from such unique, yet equally rich cultural backgrounds, with so much to share in wisdom and perspective.

From the fire, we adjourned to the grounds of the Egyptian Museum, where a party ensued for all attendees. DJ Eric Terena was on the decks, resplendent in his traditional Amazonian attire, the music a heady fusion of techno drum beats, goa trance and shamanic chants.

On the dance floor, I bumped into many people I had previously encountered on my journeys and we talked about our friends and neighbours, the trees. I asked each person why they were so dedicated to our forests, and in a variety of languages I received the same answer and expression – as if to say, 'Come on, you already know the answer.'

And of course, by now, I did.

'We are the trees.'

I realized then that my journey around the world in search of trees was a journey into ourselves. An exploration of our own attitudes and beliefs and a voyage of discovery into the concept of respect.

The true story of environmentalism does not take place around the bickering of policies in the Houses of Parliament, or inside boardrooms of government agencies discussing the future of far-flung ecosystems. It takes place in isolated regions of verdant forests, where communities have lived for thousands of years as custodians, in fine equilibrium with their natural surroundings.

In the sacred Maya text, the Popol Vuh, humans do not domesticate or sentimentalize animals or plants, but recognize their forms of existence as distinct and inherently valuable. Animals are not viewed simply as creatures to be used, but as essential contributors to a larger cosmic conversation, providing insight, guidance and companionship to humans, not lower beings, as many of the Europeans believed, over whom God had granted man dominion.

For the K'iche' Mayas, animals and trees were neighbours, alter-egos and even vessels of communication with the gods. Their perspective nurtures mutual respect and paints a portrait of a shared life that blends the spiritual with the material, the human with the animal and the vegetable. The natural world is perceived as teeming with messages and meaning, highlighting a symbiotic relationship that stands in stark contrast to the exploitative practices common in European, dominion-focused cultures.

The approach of the Mayas is a holistic one that indicates a deep understanding of ecological balance and spiritual connectedness. It is the type of approach that modern humans could learn from if we ever hope to chart a new course that honours the dignity and autonomy of all life forms, rather than elevating the supremacy of man at the expense of other living beings.

In an epoch marked by the rapid proliferation of urban sprawl and fervid industrialization, indigenous communities stand as guardians of endangered wisdom and stewards of the natural world. Embracing an entwined collaboration with nature, these communities hold ancient knowledge and attitudes that appear new to Western science, but have long-standing origins. Passed down through lineages, their mastery of the land, flora and fauna holds the epiphanies necessary for a planet amid an environmental crisis.

Despite only making up 5 per cent of the world's population, indigenous peoples collectively safeguard 80 per cent of the remaining biodiversity on Earth and 50 per cent of all enduring intact forest ecosystems.[1] These forests are some of the largest terrestrial stores of carbon.

Intuitive ways of harvesting forest resources are the reason for their success. They allow mother trees to remain intact, supplying essential nutrients to surrounding saplings until they mature. Carbon loss from unprotected forests is almost entirely from industrial deforestation, whereby intact ecosystems are flattened through slash and burn tactics.

Conversely, 82 per cent of the carbon that is released from indigenous lands is done as a result of these selective logging and garden forestry methods, notwithstanding unavoidable natural interventions such as wildfires and forest warming from climate change.[2] Should a carbon atom reside on non-indigenous land, it is up to six times more likely to be released back into the atmosphere than if on land governed by indigenous practices – even if this land is a protected area.

The Amazon basin holds some of the highest densities of indigenous lands and is still largely a frontier yet to be fully understood by Western science. Its rainforests are home to 10 per cent of recorded global biodiversity despite covering only 5 per cent of our planet's surface, encompassing the largest remaining tropical rainforest. Its diversity, natural and cultural, is immense; housing over 3 million species discovered so far and over 350 different ethnic groups, over 60 of which remain largely isolated from mainstream society.[3]

It is a tantalizingly mysterious, yet dangerous region that few truly understand. The Amazon faces many grappling hands, from those organizations trying to exploit its richness for oil, wood and pasture to those desperately trying to save it, yet it is only truly understood by those communities who have grown with the forest itself.

I cannot do justice to their story as they can, but I have had the privilege of spending time with some people who call it home, like the Asháninka, Yawanawá and Desana tribes of the Brazilian Amazon. These defenders of the rainforest have been in an existential battle against the threats of deforestation from drug traffickers, gold miners, oil tycoons and global warming that have endangered the delicate equilibrium of life in these regions since they began. Yet despite their long-standing protective efforts, these communities are only just starting to receive recognition as the true unsung heroes of conservation, who continue to fight to live in symbiosis with the forest.

Awareness is coming, indigenous voices are being raised on the global stage and their wisdom is beginning to be recognized, but there is still work to be done to elevate these communities further. After all, those who reside in the forest will always be its most suited custodians.

The loss of indigenous stewardship began with colonization. Since the 15th century, European powers have driven the erosion of indigenous communities worldwide. With an insatiable quest to extract resources like timber and minerals, as well as occupy and remould land for agriculture, colonizers enforced the displacement of indigenous peoples from their ancestral homes. This fuelled a systematic dispossession of land from these cultural groups.

On 12 October 2020, Columbus Day, a coalition of indigenous people released a very simple statement: 'We were not discovered.'

Columbus Day has been a federal holiday in the United States since 1971, when the explorer was celebrated for his part in opening up the tropics to trade. Columbus and many of the early Spanish and Portuguese navigators have long been venerated as fearless pioneers in the 'Age of Exploration' and are credited with finding mineral wealth and new lands that were ripe for the taking – boosting the global economic and political standing of expanding European empires. It is true that history is written by the victors, but thankfully we are now becoming more aware of the other side of the story.

Colonial policy by its nature, particularly in countries such as Brazil and Mexico, aimed to erase those things such as language, cultural practice and traditions that were intimately tied with the indigenous connection to ancestral land, in the name of monopolizing resources, predominantly for international trade. Enforcing foreign systems of governance and ownership, colonizers would seize indigenous lands while criminalizing the very practices that sustained these communities across generations.

Colonization of delicate biomes, such as the Amazon rainforest, continues through various means, including INCRA (the Brazilian National Institute for Colonization and Agrarian Reform) settlements. These were initially intended to promote population distribution and agricultural development. However, they have resulted in significant deforestation, with some areas experiencing up to 70 per cent forest loss. Often unfamiliar with sustainable farming practices suited to the Amazon, settlers cleared large areas of forest, and the lack of technical assistance exacerbated this problem. Many of these settlements were eventually abandoned, leading to the establishment of new settlements and further deforestation.[4]

The INCRA scheme is not the only insidious threat to its precious ecosystems. Prior to the inauguration of President Bolsonaro in 2019, Brazil had a highly regarded reputation in the world of conservation, providing significant protection to its indigenous populations and trees under numerous presidencies, including a stint in 2008 by Luiz Inácio Lula da Silva (affectionately dubbed 'Lula'). In 2004, there was a staggering loss of 28,000 square km (roughly 7 million acres) of Amazonian forest, yet by 2012, that figure had fallen to just 4,600 square km.

This somewhat peaceful balance was not to last after Bolsonaro steamed into power in 2019. What followed were some of the most devastating years that Brazil's Amazon had seen in the modern age. In the three years of Bolsonaro's reign, more than 34,000 square km (8.4 million acres), an area larger than Belgium, disappeared from the Amazon, not counting the devastating wildfires that tore through the jungles – a 52 per cent increase to the previous three years. Slashing environmental forces, he cut spending on science and environmental agencies, fired environmental experts, and pushed to weaken indigenous land rights.

Yet, January 2023 saw hope spring one more time as President Lula was reinstated in office. At a time when protections for the Amazon were looking as dire as ever, six months into his presidency there was clear-cut evidence that maybe things were brightening up: losses for the first six months of 2023 were the lowest since 2019.[5]

At the time of writing this book, Lula is grappling to rectify the damages done by the Bolsonaro administration, such as the environmental research budget cuts that threaten to leave labs literally in the dark.[6] However, for now it is a sigh of relief from environmentalists and the indigenous communities alike.

Forests contain a quarter of the world's languages. The Amazon alone holds 300 and in New Guinea there are over 800 languages. The acclaimed ethno-botanist Wade Davis calls each one an 'old growth of the mind', referring to the rich diversity of culture that has evolved in splendid isolation.

In the Congo, Batwa pygmies remain forest specialists and spiritual intermediaries, even after thousands of years of contact with farmer neighbours. Russian indigenous cultures claim their ancestry back to Siberian tigers. In Canada and Alaska, dozens of native tribes sustain old ties to the forest landscapes of the boreal. Forests have offered people socio-ecological niches in which to stay different, buffering them from the homogenizing traffic of global ideas and colonial domination.[7]

The key roles of indigenous communities and their appreciation for the delicate balance necessary to live on our planet make them crucial guides for maintaining pathways of life that prevent the overburdening of the planet.

Eduardo Kohn, an associate professor of anthropology and author of *How Forests Think*, advocates for a shift from a human-centric perspective of nature, intelligence and language. In his book, Kohn examines the interconnectedness of human and non-human entities through the lens of Amazonian cultures.

He details his time spent living with the Ecuadorian Amazon, which extended beyond human interactions to include the other beings and elements that are integral to their lives. This is what inspired him to coin the phrase 'an anthropology beyond human', reflecting the Runa's way of life, which involves hunting and cultivating in preference to purchasing food. Kohn examined how they engaged with their surroundings; their belief that non-human living beings, and even certain elements of the environment, retain a form of cognition and thought.

Kohn details the concept of 'semiotic entanglement', meaning human and non-human lives are intertwined within a web of communication that blurs traditional boundaries between them. This theory challenges conventional perceptions of self, which naturally dominate with the isolated human identity. He explains how the Runa challenge the sense of self, and perceive it as a 'we', or an 'ecology of selves', as opposed to a mentally isolated 'I' or a single unit of self.[8]

Kohn's work is a contribution to a larger academic movement that seeks to redefine the relationships between humans and the environment. A movement that emphasizes decentralizing human-centric views of the natural world and encourages a broad sense of understanding that life on Earth means all life, not just human life. Kohn's close ties with numerous Amazonian tribes, from the Sápara Nation to the Kichwa Nation of Sarayaku, have given him an intimate understanding of the principles they share. His collaboration with Sápara Nation leader Manari Ushigua is part of his ongoing effort to develop these ideas further, as seen in his 2014 book, *A Reflection on How Forests Think*.[9]

Kohn also played a role in the creation of the 'Living Forest Declaration', a visionary proposal initiated by the Sarayaku people.

According to the Sarayaku, the declaration comes from the rainforest itself, which it describes as a living entity that constitutes all life inside, not just a space for the exploitation of resources.[10] It contrasts with Western perspectives that often regard natural spaces as empty and resource-laden, suggesting instead that the so-called 'natural' world is composed of living beings engaged in ongoing communication, including humans.

The American author David Haskell agrees. 'What Western science calls a forested ecosystem composed of objects is instead [to the indigenous] a place where spirits, dreams and waking reality merge. The forest, including its human inhabitants, is thus unified . . . spirits are not otherworldly ghosts from a distant heaven or hell but are the very nature of the forest, earth and grounded, connecting soil and imagination.'

It might seem bizarre to humour and entertain such ideas. And yet, as Haskell explains, 'The Western mind can perceive and understand abstractions such as ideas, rules, processes, connections and patterns. These are all invisible, yet we believe them to be as real as any object. Amazonian forest spirits are analogous, perhaps, to Western reality dreams such as money, time and nation-states.'[11]

This is an eye-opening perspective on the interconnectivity between humans and the environment, one that is often linked to the more spiritual aspects of forests and the people who live in them.

In the realm of ancient forest wisdom, it is the shaman who emerges as the spiritual interlocutor. The mere word conjures up images of a man cloaked in ethereal robes, grasping a gnarled wooden staff, his body adorned with the hides of wild animals, the teeth of predators, feathers, beads and fetishes of ancestral connection. The shaman's weathered face is etched with stories of hard journeys and a gaze that penetrates the veil of reality. For thousands, perhaps hundreds of thousands of years, the shaman has navigated between realms embodying the bridge between the physical and the spiritual, guiding seekers through the labyrinth of consciousness and unlocking the ancient knowledge that lies dormant within.

To be a shaman is to act as an intermediary between the human realm and the natural world – or to reinforce the fruitful exchange between the two. Shamanism simply encourages the act of remembrance, that we are not apart from nature, but integral to it. When the first Western explorers encountered shamans – or witchdoctors – in the late 18th and early 19th century in remote communities entering a state of trance, they assumed that they were experiencing the supernatural or exploring their inner world, but the truth is that these 'medicine men', and women, saw no distinction.

As Karen Armstrong explains in *Sacred Nature*: 'He or she was not looking beyond or above nature, nor was he seeking the divine within himself, like a modern contemplative. Instead the shaman projects his awareness outwards into the depths of the landscape, which for him is alive, spiritually, psychologically and sensually. He experiences an awareness which he and his community have in common with the animals, insects and plants around them.'[12] While we in the West might scoff at this, for our own ancestors this primordial understanding would have been the norm.

As we have learned, the forest plays an important role to indigenous communities around the world. And for many the tree is central to worshipping the divine, in whatever form it is recognized. Shamanism is the embodiment of this mind–body connection with nature; the word *shaman* originates from the Tungus tribe in Siberia and China, an indigenous group that dispersed across the Arctic and Siberian plains and whose shamanic traditions remain deeply connected to nature. While they have evolved to coexist with modern life, they have preserved their ritualistic ties to the natural world.[13]

Shamanism is now a term attributed to indigenous cultural wisdom of any lineage and describes a sense of spiritualism led by nature. In essence, it hinges on the acknowledgements and understanding that all forms of life, human and non-human, have spiritual value. It is a shaman's role to interpret and share this; often taking on the role of a medicinal healer with knowledge of plant life, or forecasting and helping to improve hunting or fishing prospects.

These medicine men and women serve as custodians of the expansive biological knowledge of generations of ancestors about the ecosystem they inhabit. Trees are entrenched in the practices of shamanism – the antennae of the Gods – serving as bridges between the human mind and the wisdom of nature.

A decade ago, I visited the coldest inhabited place on Earth – Oymyakon in Yakutia, a remote province of eastern Russia. It is a place with winter temperatures that regularly plummet to sub-60°C. And yet in the summer it can be a blazing inferno, and temperatures are rising each year. I was there to document the local reindeer herders as they embarked on their annual migration, and to visit the indigenous Yakut people, famed for their deep spirituality. Earlier that year, wildfires had ravaged the landscape – eradicating huge swathes of forest, destroying villages and melting the permafrost.

Anya, a local lady dressed head to toe in fur, told me that the fires were only stopped when the *shaman-derevo* (shaman trees) were invoked to provide protection. By making offerings and praying to the spirits of nature, balance was restored.

Yet shamanism goes beyond the spiritual guidance of others. For cultures that exist in regions like the Amazon, the practice centres around the dense forests and plant species that hold healing properties. With knowledge passed from mother to son, and father to daughter, tree shamans, or *paleros*, believe in the sacred medicines of trees. They take years of apprenticeship, where they undergo *dietas* – periods of time in the jungle when they must ingest infusions made from the barks of the trees, be on a very restricted diet, and cannot be physically touched by anyone.

Shamanism assumes various roles within different cultures, but it exists to guide people to reconnect with the natural world. As has happened throughout history, it is often the shamans and spiritual leaders of indigenous communities that are leading the fight to keep this knowledge alive, and in doing so fight against the desecration of nature.

In the forests of North America, the totem pole, with its rich iconography in the ancestry of First Nation culture in the Pacific Northwest and British Columbia, is far more than an artistic pursuit. It is a pillar of family lineage, spiritual reverence, and vital cultural heritage that spans epochs – the oldest remaining totem pole stands to be over 9,000 years old.[14]

These symbols demonstrate a deep intuitive connection to the forests in which they are harvested and embellished. They are made from the rot-resistant western red cedar, and many coastal First Nations communities perform ceremonies displaying their gratitude and respect for the sacrifice of the tree.[15] The totem trees stand as feats of cultural perseverance against some of the most prolific colonial genocides of indigenous peoples on record.[16]

The grapple for the reclamation of rights to ancestral land is at the heart of the indigenous resistance movement. It embodies the perpetual fight for the protection of ancestral territories from encroachment, exploitation and destruction by those without just cause or right to do so. In the modern age, this quest for justice often takes place in courtrooms and voracious political protests. It is a tale we know unfortunately too well, and a pattern only likely to worsen with depleting fossil fuel resources and a global lethargy in converting to green alternatives. One such example is the battle between the Standing Rock Sioux Tribe, custodians of the Standing Rock Reservation, and the Dakota Access Pipeline.

Situated across the border between North and South Dakota in the United States, the sacred land that holds the Standing Rock Reservation became subject to oil transport developments in the form of a pipeline transporting 750,000 barrels a day. Despite fierce and eventually violent protesting from the collaborative efforts of indigenous communities, the 1,200-mile-long pipeline was built – splicing the ancestral Lakota land in two and placing communities at risk of decimating oil spills and contaminated water supplies.[17] The protests that preceded construction, known as the Standing Rock protests, became symbolic of indigenous resistance and the fight for land and environmental justice.

The teachings of the indigenous peoples of the Americas are some of the most poignantly centred around our Earth, not in a fantastical biblical way, but in a humble cherishing of all aspects of our natural world. As relayed by the Pawnee Eagle Chief Letakots-Lesa:

Tirawa, the one Above, did not speak directly to humans . . . he
showed himself through the beasts, and from them and from the
stars, the sun, and the moon should humans learn.

A few hundred kilometres away flying directly over the North Pole, on
the European fringes of Arctic Circle, where frigid temperatures and
rarified sunlight reign over the boreal forest, the people of the reindeer
god reside. The Sami once inhabited a region stretching from the
Lapland regions of northern Scandinavia to the Kola Peninsula of
Russia (known by the Sami as the *Sápmi*) for thousands of years.
Since 100 CE, and the arrival of the Finns into Finland, the Sami have
been dispersed from their traditional lands, along with the near-
extinction of their native tongue.

Once nomadic, the Samis' close bond with reindeer is tantamount
to their cultural survival, and their intimate knowledge of pastures
and tundra has been cultivated over their thousands of years of
coexistence with the environment.

Their observations, interpretations and reflections on prolonged,
multigenerational ecosystem change of their surroundings enrich
over time. Yet, despite being ignored for decades, the 100,000 Sami[18]
that remain are steadfast in voicing the urgency of the acute
degradation of the landscapes in which they have coexisted for
millennia. For their efforts, the Sami are the only indigenous group in
Europe to be recognized by the United Nations.

Jannie Staffansson, a Sami reindeer herder, in essence is raising
the voices of the Sami on the global front of climate change. Armed
with a degree in organic chemistry, Jannie consulted on the Paris
Agreement and the United Nations Framework Convention on

Climate Change, bringing awareness and demanding the native community of the Arctic be taken into consideration.

'Climate change was a commonplace topic of discussion in my indigenous community,' shares Jannie, speaking on her reasons for seeking a Western education. 'I told my father to go and tell the rest of the world about this, but he said, "Jannie, you go and do that. Get yourself an education and then they might listen to you. They're not listening to us." I realized then what I had to do.

'I feel very strongly that the mainstream society is tackling this huge crisis in nature backward,' says Jannie. 'World leaders and researchers know that indigenous peoples hold a lot of knowledge, and together we could come up with better solutions and make better decisions, but they are not interested because we have been colonized and stereotyped as being less and knowing less.'

Sadly, it is this mindset that has got us into the place we are today. The loss of Sami culture comes with a huge and unexpected price. The Sami have herded reindeer for at least 2,000 years, probably far longer. Rock art in Lapland shows stick figures of humans and reindeers that are 8,000 years old. Their relationship has been one of love and interdependency for millennia. The animals allowed the Sami to survive in an unforgiving world of cold and ice that would mean death to anyone not wearing clothes and shoes made of their skin.[19]

'Reindeer are life. They are everything. Without reindeer we die,' said one Sami leader. Their lifestyle is under threat. For decades, the Sami way of life was eroded by the Norwegian, Swedish and Finnish governments' collective efforts to 'assimilate' the natives in the 20th century. The Sami were forced to accept Christianity and abandon their traditional shamanist worship, their lands were stolen by successive governments, and in Sweden the Sami were even subject to forced sterilization and a eugenics programme.

If all that was not horrific enough, they now bear the brunt of climate change. Warmer winters are deadly for the reindeer, as their food sources face irrevocable changes and the winter ecosystem collapses. But it is not only the Sami and their reindeer that suffer – we all do.[20]

Reindeer like to eat the shoots of new growth trees and keep the birch and larch forests in check to ensure the treeline does not encroach too far north. Surely lots of trees is a good thing, right? Well, not in the Arctic regions, as I learned in Greenland, because as the treeline spreads northwards it heats the ambient temperature of the ground, resulting in the defrosting of the permafrost all across the Arctic regions.

This in itself is catastrophic, because stored beneath all the ice is billions of tonnes of ancient methane, which is around 80 times more toxic to the atmosphere than CO_2. If that is allowed to leak, then global warming will reach its tipping point much sooner than anyone expected. We might do well to listen to the Sami voices and encourage a return to indigenous practice. Reindeers are not just for Christmas after all.

Indigenous wisdom comes from a deep history with the land that communities inhabit. It would not come as a surprise, then, to assume that the real Garden of Eden lies in a land with indigenous ancestral ties. Some say it is nestled amid the baobab trees of a small group of raised islands on the Makgadikgadi salt pan of Botswana. On what used to be an enormous lake, these beacons of isolated beauty have been proven by DNA analysis as the source of modern humanity. Scientists recently analysed mitochondrial DNA samples from hundreds of people across Africa and ran computer simulations to determine where they came from. Around 200,000 years ago, *Homo sapiens* set off on a new journey of discovery.

I visited Kubu island in the Makgadikgadi in 2019 while following elephant migrations across the plains and my guide Kane showed me the places that his ancestors called home. All across Botswana and Namibia we discovered ancient petroglyphs and rock art, some dating back to before the last ice age. Kane was proud of his San heritage, the bushmen whose culture is perhaps the oldest in continuation in the world today.

The semi-nomadic San have lived as hunter-gatherers all across southern Africa for the last 200,000 years and for much of that time they remained isolated from other African ethnic groups who spread further afield. Their culture retains some of the most precious ecological knowledge anywhere on Earth.

In one of the most remarkable encounters I have had the privilege of experiencing, Kane led me to the site of a lion kill near to our camp one day. He had followed the tracks of four adult male lions, who were in the process of killing a buffalo. Kane, armed only with a short spear (he refused to carry a gun unless the law required him to do so), encouraged me to follow him to hide behind a bush around ten metres away from the lions and watch as the buffalo took his final breaths. The lions began to devour it from behind, guts first, and then after an hour or so, he turned and said to me, 'This is how my mother got our meat when I was a baby tied to her back.'

At which point Kane stood up and came face to face with the lions. He invoked a prayer to his ancestors and stared the lions down. After a tense few seconds they retreated to a nearby acacia tree five metres or so away. 'Now, follow me,' he ordered. I stood up, despite feeling utterly terrified at the prospect of walking straight towards the wild lions. Kane boldly approached the dead buffalo and showed me how his mother would take a knife and cut away some of the meat to take home. The lions simply watched on from the shade and let us do what we needed to do.

'We never take more than we need,' he said with a huge grin. 'There is plenty of meat to go round for all of us.' And with that we backed away from the kill and let the lions return to finish their lunch.

To the north of the Kalahari lies another so-called 'wilderness'. Second only to the Amazon as the largest tropical rainforest ecosystem on the planet,[21] the Congo basin is home to immense cultural diversity, inhabited by not only 10,000 species of animals and 600 different types of trees, but also 150 distinct ethnic groups.

One of these groups is the Batwa, one of the last remaining hunter-gatherer societies in existence, who inhabit the 'impenetrable' Bwindi National Park. The Batwa (or *Twa*) have lived side-by-side in this wilderness with their fellow occupants for an estimated 60,000 years, even surviving the depredations of the enslavement and colonial eras.[22] Until relatively recently, the Batwa populations have managed to preserve their rich culture and traditional way of life.

Living off forest game, such as antelope and guinea fowl, the Batwa once spread across the vast forested regions of the Congo basin, but now are forced into rapidly shrinking areas of the forest by other groups. Neighbouring agriculturalists, who deforest without restraint, discriminate against the Batwa in hopes of forcing them further from their settlements. However, the Batwa face dispossession of their land and loss of stewardship over the forest not only from those who wish to convert it to pasture, but also from those with good intentions rooted in conservation.

The Batwa have shaped and sustainably occupied the Kahuzi-Biega forest for millennia, acting to safeguard and engineer livelihoods in one of the most biodiverse wildernesses on the planet. But then, 'They came to purge the forest by force.'[23]

As has happened across central Africa, the Batwa were forcibly displaced from the Kahuzi-Biega and their homes burned to ash, leaving their community to rot on the outskirts of the forest, where they suffered severe discrimination from neighbouring communities and no access to healthcare. Attempts to reclaim their ancestral land are often met with belligerent force, violence and death at the hands of internationally funded and trained park rangers, who are instructed to remove local communities as a way of protecting wildlife against poaching.

There is a fine line to tread when it comes to habitat protection, conservation and indigenous rights.

The indigenous communities of Australia are some of the oldest cultures surviving today, with myth and lore that stretch back some 65,000 years – orally passed between lineages. In his seminal work *The Songlines*,[24] Bruce Chatwin dived into the sheer complexities of the oral history of the Indigenous Australians before the arrival of European colonizers, and how its roots traverse the globe.

The Songlines delineated the tracks laid out across the Australian landscape and beyond, which the indigenous groups used as navigation, storytelling and a means of memorializing vital information about the landscape – the 'songlines'. Indigenous peoples believe these tracks were laid out in the 'Dreamtime', when creation and natural forces merged to create the landscape. Chatwin detailed how the main songlines across Australia enter the country from the north or north-west, from across the Timor Sea or the Torres Strait, proceeding to weave southwards and across the rest of the continent. It is a common belief that these songlines represent the routes that the first Australians took as they traversed the continent upon arrival all those years ago.

In Aboriginal culture, everything on Earth is held together by the songlines and everything is subordinate to dreaming, which is constant but ever-changing. Every landmark is wedded to a memory of its origins, and yet is always in the process of creation. Every animal and every tree pulsates with ancient energy and yet is still being dreamed into existence. The land therefore is encoded with everything that ever has been and everything that ever will be.[25] Perhaps the Aboriginals conceived of the concept of DNA long before we in the West did. To walk the land is in itself to engage in a constant act of worship, of endless creation.

Sadly the Europeans who landed ashore in the 18th century saw only primitive tribes. They lacked the imagination to even begin to understand the spiritual comprehension of the Aboriginals. All they saw was a people with no metal tools or written language familiar to Europeans, a people who did not farm or cultivate the land. These people, who generated no surpluses and whose imprint on the landscape was minimal, were seen as uncivilized, incapable of little more than starting a fire.

What the colonizers did not appreciate was that cultural burning – or 'fire stick' burning, where small, low-intensity fires are carefully introduced across the landscape in a mosaic – actually reduces the amount of dry tinder present around the bush. This decreases the potential intensity of wildfires that often spark late in the dry season.

Among other practices, this is the product of multigenerational, orally passed-down wisdom that protects and nourishes the ecosystem in which the indigenous populations are so closely intertwined.[26]

When the British reached the shores of Australia, they were incapable of embracing the wonder of Aboriginal culture. It was inconceivable that a people could choose such a way of life. Aboriginal Australians had a profound and intricate connection with their land. But progress and improvement over time were the hallmarks of the Victorian ideal of development, and to the European eyes the Aborigines were the embodiment of savagery.

The behaviour of the colonizers represents one of the darkest chapters in the British imperial story. Aboriginals were regarded as no better than beasts; they were shot for sport, their bodies hung as scarecrows, and they had no legal custody of their own children. One man of God, a Reverend William Yates, said, 'They were nothing better than dogs.' Their doom, wrote Anthony Trollope in 1870, 'is to be exterminated, the sooner the better.' As late as the 1960s, Aborigines were included in a school textbook as one of the more 'interesting animals of the country'.

These settlers were ignorant, brutal and offensive towards the native people, who for millennia had thrived as hunter-gatherers and guardians of nature. As Wade Davis says in *The Wayfinders*:

In all of their tenure a desire to tame the rhythm of the wild have never touched them. The Aborigines accepted life as it was, a cosmological whole, the unchanging creation of the first dawn, when earth and sky separated and the original ancestor, the Rainbow Serpent, brought into being all the primordial ancestors who, through their thoughts, dreams and journeys, sang the world into existence.[27]

The colonization of Australia saw the land proclaimed as *terra nullius* (meaning 'land belonging to no one'), rejecting the nomadic population's claim to their home. Separating these groups from the land in which they 'sung into life' became the ultimate form of oppression. As Chatwin touches on, this began to erode the sense of belonging in these communities, deeply threatening the future of the stories and songlines, the tracks of communication – and therefore the identities of future indigenous groups. By the early 20th century, a combination of disease, exploitation and murder had reduced the Aboriginal population of Australia from over a million to a mere 30,000. The land had been transformed from Eden to Armageddon.[28]

Yet, it is in the songlines of Indigenous Australians to adapt to the world around them, rather than force the world to change to meet their desires. While it is chilling to think that in the 19th and early 20th century a huge repository of wisdom was lost, a few vocal leaders fought to preserve a way of life.

Thankfully, times changed, and the culture was not completely eradicated, but it was not until the 1980s that Aboriginal rights started to be respected and their culture finally honoured. Sadly, there is still a lot of discrimination.

Fighting against any prejudice that still exists, listening to indigenous communities and heeding their wisdom is going to be the best way to improve our lives and the health of our planet.

Chapter 11

Healing Forests

————

In the woods, we return to reason and faith.

–Ralph Waldo Emerson, *Nature*

Dappled afternoon light cast soft shadows onto the forest floor, framing a clearing on the banks of the river. A community of Asháninka people were beginning to gather for a ceremony later in the evening. They had just finished their daily tree planting. Their leader, a 45-year-old man by the name of Benki Piyãko, sat perched in a hammock in a T-shirt and shorts, grubby from the soil. His face is weathered from a hard life outdoors, but is kind and exudes a wisdom beyond his years.

For centuries, the Asháninka people have lived relatively isolated lives, existing in harmony with their environment by fishing, hunting and gathering fruit. Their home rests on the border between Brazil and Peru, in the heart of the Amazon rainforest. Wildlife is truly abundant here, but it is not designated as a conservation area or even protected. Life thrives because of the Asháninka people. Over decades, they have restored and nourished the land while sustaining a thriving community. 'Our relationship with the forest is based on respect,' said Benki. 'If we must cut down a tree, we must replace that tree. The forest looks after us, so we must look after the forest.'

The Asháninka, like other indigenous communities of the rainforests within the Amazon basin, are some of the world's best conservationists. Their profound knowledge of the ecosystem allows them to master sustainable cultivation practices that provide food security without compromising the integrity of the tropical rainforest ecosystem. This is typically agroforestry, or 'simultaneous polycultures',[1] which involves the integration of numerous crop species with a diverse array of tree species native to the surrounding area.

The Asháninka have long been stewards of this wilderness, but in recent years the dark force of deforestation has cast a deadly shadow over their delicate balance of life. A mere mile or so away, their land is being decimated by illegal logging activity in the name of mineral mining and exploitation. Every day this activity inches closer to their settlement, like a fire spreading through a home. If, or when, the loggers reach their settlement, the rich and flourishing climate they have built over lineages will be wrecked, along with their culture, way of life, and intimate knowledge of the forest.

This only fuels the fire in Benki. I sat and listened to his sermon, translated from his native tongue, first into Portuguese and then into English. 'The great deal of richness that exists here, from the forests, animals and plants all thrive still as a direct result of the way we have guarded and tended to this area since 1986, when we began our work.' His brow furrowed. 'We maintain our culture well, and protectively, but for all that we protect we carry great worry.' He broke away, looking up as something rustled up above in the canopy.

'Deforestation is one of the greatest threats we fear, it is happening all around us. People felled our forests, and that made our rivers very dry. We now have unbearable heatwaves, rains during the summer time as if it were winter time, and dryness during the rainy season. Our worry about this destruction is what is driving us to take our message, as indigenous peoples, to the whole world. Our environment is the security of our lives. If we don't take care of all these species, of this richness of nature, we are heading towards a great catastrophe that may affect us in a very deep way.'

Rudy Randa first came to the Amazon in 2013 on a spiritual retreat to escape the hedonism of Los Angeles. Since then, he has returned many times to deepen his understanding of indigenous knowledge and to spread this wisdom across the Western world.

We met in Benki's village in the roundhouse, a large thatched hut used for communal meetings. Rudy had facilitated the gathering under the auspices of an organization called Aniwa, which brings groups of Westerners to the Amazon to participate in traditional plant medicine ceremonies. These retreats are typically ten days long and involve nine hours of meditation and singing, all the while consuming the entheogenic brew Ayahuasca.

'I guess you could call it a spiritual awakening. It certainly changed, perhaps even saved my life,' he told me. 'I was feeling low, had lost my way and somehow found myself in the jungle. It was there I had this download that told me to live a simpler life, and to come here and be a bridge between cultures. I fell in love with the Amazon and the way of life here in Brazil.'

But Rudy is no cliched new-age hippy. He walks the walk. In 2015, after meeting with the leader of the Huni Kuin tribe, Ninawa Pai Da Mata, he established his own conservation charity, the BOA Foundation. This works in alliance with indigenous communities to preserve and protect sacred land by support projects such as strategic land buybacks, the restoration of native ecosystems, and fundraising for sustainable living solutions, tree planting and access to clean water.

He spends at least three months a year in the Amazon helping the Asháninka and other communities to plant trees and his foundation has raised over $3 million to help fight illegal loggers and stop cocaine trafficking.

'I genuinely believe that the quest to restore the Earth starts right here,' he said. 'By bringing Westerners out here to learn more about the culture, to sit with the plant medicine and have transformative experiences, it will have a ripple effect that brings awareness to the world.'

I was starting to see that myself. Small groups of Westerners were sitting around campfires discussing their own projects. A group of Hassidic Jews from Brooklyn had come to deepen their own religious beliefs, and saw no reason that their Judaism was at odds with the shamanistic experience. I met Ryan, an Australian musician who was recording an album with some big name artists; Robert, a tax accountant from London; Mike, a property tycoon from New York and Jacob, a watch dealer living in Mexico. Sara, his girlfriend, was an artist from Rio de Janeiro and Vivien, a local Brazilian too, had come to share lessons from her own trees. All of them had grand plans to go on and do something to help save the forests as a result of their recent experiences.

Rudy is a man that truly bridges the divide. In LA he mingles with Hollywood stars and high-flying businessmen, organizing conferences and dealings with music labels, all the while helping to promote Amazonian culture to the West. He has flown leaders like Benki and Ninewa to New York, Davos and London, to give them a platform to speak about their way of life to a wide audience.

'I believe that traditional culture can go hand in hand with modern technology. Look, these villages have solar power and Wi-Fi if they want it. But their impact on the world is minimal. They have a true spiritual connection to the planet, they're happier and have less fear. And all of this is down to their reverence for trees. As Benki says, if every person on this planet planted or cared for just one tree, we wouldn't have an environmental crisis.'

The entire village congregated in the roundhouse. Men, women and children sat in the pitch black on hard wooden benches. They had fasted and undergone a rigorous diet. They had washed and cleansed themselves in preparation for the ceremony. Some of the Westerners wriggled around in discomfort, unsure what the Ayahuasca would bring. We had all heard the stories: purging, vomiting and experiencing the sensation of death. We knew that the *Banisteriopsis caapi* vine has been used in the Amazon for thousands of years as a form of medicine – a cure for past traumas, emotional attachments and physical ailments alike. For the Asháninka, it is equally a ritual, a religion and a rite of passage. For me, it was an awakening.

Benki, now transformed from tree planter into high priest, wore a long white robe and a bandana of embroidered cloth around his head. All of the villagers were barefoot and their faces were covered in red ochre paint. The ceremony began, as one by one we walked up to the altar and were presented with a cup of sticky brown liquid that looked, and tasted, like sweet tar filled with the detritus of vegetable matter. After consuming the potion, we sat back and waited.

There was a long silence, and then after an incalculable time Benki led the chanting. An impossible symphony of wild screeches, deep groans and high-pitched wails. Sometimes the women led, then it was a child. To an untrained ear, there was no rhyme nor reason to the noises, until I came to realize that the people were simply singing the song of the forest. It is a song that harks back to the beginning of time, invoking the spirits of the birds, animals and trees to come alive. And soon enough, come alive they most certainly do.

As I closed my eyes, shapes began to appear, beautiful fractal visions that seemed to echo the weave of the cosmos itself. Flashes of vibrant light streamed into my consciousness and I felt awake as if for the first time in my life. Then came the snakes, great anacondas weaving like vines through every inch of my being. I felt every emotion possible, and some I never thought possible at all. There was terror and desperation. I went back in time and re-lived every part of my life – the good, the bad and the downright ugly, through vision, smell and feeling. I recalled being born, and I remembered what happened before that too.

There were colours I had never experienced before, and I could see the music dancing around me. At one point I became a bee. I lived its entire blissful life from start to finish and began to appreciate what an inherently wonderful existence it lived – free from ego, free from fear and free from doubt. For a fleeting moment that felt like a magnificent eternity, I was the vine, I became the tree. I felt the motion of chlorophyll passing through my cells as I really, and I mean really *felt* photosynthesis. And now I understood. In a great white vastness I was at one with all and everything, lost and found while at the same time wrapped in a blanket of sheer bliss and love.

The chanting and music eventually ended and all I heard was morning birdsong. Benki was still sitting on his bench, both solemn and joyous. '*Só Alegria,*' he proclaimed. 'Only joy.'

He is right.

Medicine? A wild trip? Just a mind-altering substance, you might say? Perhaps. And I am well aware that to the uninitiated, all of the above sounds utterly insane. Maybe it is. But now I have no doubt whatsoever. 'We are the trees.'

Since the dawn of the collective intensification of agriculture some 10,000 years ago, our planet has begun a journey of drastic devolution towards desolation. According to experts, we are facing the commencement of the sixth mass-extinction event[2] that our planet has witnessed, and the first suspected to be caused by the activity of human life. Our unsustainable use of land, water and fossil fuels is fanning the flames of climate change.

It is hard to ignore the cultural division of blame for the crises we are in. In his striking cosmo-ecological manifesto *The Falling Sky*, Amazonian Yanomami shaman Davi Kopenawa details how deforestation foretells the termination of life: 'The Earth's skin is beautiful and sweet smelling. The white people only know how to abuse and spoil the forest. They destroy everything in it, the earth, the trees, the hills, the rivers until they have made its ground bare and blazing hot. All that remains is a soil that has lost its breath of life.'

How have we become so dissociated from the natural world? It is not only the Western focus on monetary gain or expansion of power, but how our science mimics these emotional voids in our attitude to the natural. Western categorization of nature is inherently hierarchical – a quality that many indigenous cultures simply avoid. As written by Robin Wall Kimmerer, in her series of essays *Braiding Sweetgrass*, Western tradition and society place plants at the bottom of our hierarchy of beings, reduced to means of fuel; and in the same vein, how little humans understand how to live on Earth compared to other species that have long out-existed us already.

This calls into question what Western science and ways of living fail to acknowledge: we must look to the teachers among other species for guidance. The complex forestry systems – that demonstrate elite resourcefulness, intelligence and interconnectedness with other species – make the case for plant ethics based not on a state of 'otherness', but a respect and desire to learn, akin to seeking guidance from a wise elder. It is when considering this gap in Western attitude that it is vital to recognize that this way of thinking owes its roots to indigenous cosmologies and perspectives.[3]

A growing body of scientific evidence supports the notion that indigenous intervention in forest ecosystems can leave these sites in a healthier state than if they were simply left untouched. These discoveries all began with a hazelnut tree. Growing far from their native range, lost within a sea of gigantic cedars, these seemingly conspicuous trees caught the attention of First Nation leaders who summoned ethnobotanists to investigate.

What the researchers and leaders quickly realized was that they were looking at a flourishing, human-cultivated ecosystem. The leaders began to recall histories of long-forgotten gardens, whereby their ancestors had cultivated edible and medicinal plants like hazelnut trees within the conifer forests.[4] Upon reflection, the researchers found that these sites where indigenous ancestors had intervened harboured strengths that the surrounding 'natural' conifer forests lacked. Healthier, more biodiverse insect and animal populations, more effective sequestration of carbon, and greater resilience to events like drought – all while providing food security for the ancestral community.

On the island of New Guinea, 90 per cent of the forest is held by native communities. Across the Pacific, Brazil's 1988 constitution established the country as a world leader in the legal recognition of indigenous peoples' rights to their ancestral territories. In Colombia too, the government enshrined similar protections. Millions of acres of aboriginal forests are now protected, and these are the healthiest forests of all.[5]

This shows us how indigenous people can often outperform the efforts of Western environmental agencies and conservation organizations at supporting biodiversity and fortifying general ecosystem health. Scientists are finding that leaving nature alone may not always be the best course, and that handing these environments over to the original land stewards may be a crucial step in improving conservation efforts. Moreover, on a fundamental level, these findings show us how, as humans, we are capable of empowering nature, not just draining it.

Many old Japanese folk stories revolve around the Kodama, a kind of spirit or deity that lives in the trees. People believed that Kodama travel around the forest, retaining ancient knowledge that is passed down through the generations. If you cut down a tree that has a Kodama living in it, you will be cursed.

The most ancient and revered tree in Japan is the Jōmon Sugi, a large *Cryptomeria*, or Japanese cedar, on the southern island of Yakushima. Thick with moss, ferns and often shrouded in mist, the forest exudes a fairy-tale-like energy. Fittingly, this place was the inspiration for Hayao Miyazaki's anime film *Princess Mononoke*, which features the mythical Kodama and the epic struggle between mankind and nature.

The Jōmon Sugi is hollow at its centre and so it is impossible to date it accurately by counting the rings, but some scientists have suggested that it might be as much as 7,000 years old, which would make it the oldest singular living tree on Earth. While its age is in dispute, what is not is the importance that trees play in Japanese culture.

Both of Japan's official religions, Shinto and Buddhism, believe that the forest is the realm of the divine. For Zen Buddhists, scripture is written in the landscape.[6] The natural world itself is the word of God. In Shinto, the spirits are in the trees, in the rocks, in the wind and in the rivers.

Nature is not separate from mankind as it is by Western definitions. The need to keep harmony between the two can be seen in every aspect of Japanese life, from the design of many homes to the affection given to gardens and bonsai trees.

Shizen, which translates as nature, is one of the seven principles of Zen aesthetics. It reminds us that we are all connected to nature spiritually and physically, and the more closely something relates to nature, the more beautiful it is. Japanese art often portrays natural scenes where trees, mountains or waves are the dominant subjects, and humans play a minor role.

The 1980s hailed an economic boom in Japan. In the opulence of the times, Tokyo businessmen were known to carry around gold flakes to sprinkle on their food and in their drinks. Money was fast and fluid. It seems apt that also at this time, a concept developed that was a counterbalance to the capitalist frenzy, a panacea to the stress, speed, overwork and anxiety of everyday life.

In 1982, in a nod to traditional Shinto and Buddhist practices that revere nature, Tomohide Akiyama, the director of the Japanese Ministry of Agriculture, coined the term *shinrin-yoku*, or 'forest bathing'. This practice of forest immersion is an invitation to heal through nature. Participants disconnect from modern devices and remove other distractions to reset within the therapeutic forest environment.

With this new field of study, the government started to test whether the forest environment had positive effects on blood pressure, heart rate, cortisol levels and immune system responses. Evidence came back supporting what was intuitively suspected. In one study, subjects were exposed to three scents commonly found in Japanese woods – cedar, hiba oil and Taiwan cypress – and all the participants experienced stimulated activity in the prefrontal cortex of their brains, which allowed for increased focus and concentration and a greater degree of relaxation.

The source of these benefits has been traced to the volatile secondary compound *phytoncide*, which trees and other plants emit when repelling insects and other predatory organisms. Why humans should be stimulated by this is still unknown. But forest bathing works.

Dr Qing Li of the Nippon Medical School in Tokyo describes forest bathing as 'simply being in nature, connecting with it through our sense of sight, hearing, taste, smell and touch . . . when we open up our senses, we begin to connect with the natural world.'[7]

He cites other proven benefits too, including: reduced blood pressure, increased NK cells, reduced stress hormones and a balanced autonomic nervous system, as well as reduced anxiety, improved sleep, a counter to depression and even the release of anti-cancer proteins. It is almost as if we are evolved to be in tune with the forest.

With nearly half the adult UK population taking one form or another of prescribed medications, and around a quarter taking more than one medication, forest therapy offers an alternative to our struggling immune systems. According to the US Environmental Protection Agency, the average American spends 93 per cent of their time indoors. Europeans are not that much better. We spend between six and ten hours a day glued to our computer and phone screens – addicted to technology and the dopamine rushes that social media encourages.[8]

With so many benefits, it seemed remiss not to try. The first trial system for the forest-bathing practice was created in Akasawa, in Nagano prefecture. In the 1990s a series of government-sponsored 'Shinrin-yoku Trails' were established, to support citizens actively to participate in this healing. Now there are 65 such trails in Japan, each with self-guided programmes for forest immersion, as well as forest therapy guides.

The Indian poet, writer, philosopher and social reformer Rabignath Tagore (1861–1941) also agreed with this vision. He held firmly to the idea that learning should be done outside, in nature, and in the schools he founded, classes were mainly conducted under the shade of trees. It would be terribly sad if we were to say we were the last generation that played in the woods.

Forests have given humanity another great gift: phytotherapy, the use of plants for medicinal purposes, arguably the most ancient form of medicine. Forests harbour a treasure trove of plant species with potent medicinal properties, offering natural remedies for a myriad of illnesses. The bark of the Pacific yew tree, found in North American forests, yields compounds used in chemotherapy drugs to treat cancer. Similarly, the rosy periwinkle (*Catharanthus roseus*), endemic to

Madagascar's rainforests, is the source of vinblastine and vincristine – essential chemotherapeutic agents used in the treatment of childhood leukaemia and Hodgkin's disease.

However, out of the approximately 50,000 known medicinal plant species, which serve as the foundation for over 50 per cent of all modern medications, up to a fifth are under threat of extinction at various levels – local, national, regional, or global – due to deforestation. And yet, our knowledge about potential drugs that can be extracted from rainforests remains in its infancy.

Knowledge of these medicines is the legacy of generations of indigenous communities using the forest as their pharmacy. When we destroy forests, we run the risk of not only species endangerment and extinction, but the obliteration of medicines we have yet to identify. Possible remedies for cancer, heart disease and diabetes are growing amid the trees, just waiting for our senses to grow sharper. Not only this, but it is estimated that around 80 per cent of the world's population living in the developing world relies on traditional plant-based medicine for primary healthcare.[9]

Humans have long relied on the vast array of botanical resources to treat ailments ranging from infections to chronic diseases. Willow bark has been used throughout the centuries in China and Europe, and continues to be used today for the treatment of pain (particularly low back pain), headache, fever, flu, muscle pain and inflammatory conditions, such as tendinitis. The property within the bark responsible for pain relief and fever reduction is a chemical called salicin, which acts like aspirin.

Moreover, forests play a crucial role in mitigating the spread of infectious diseases. Research has shown that intact forests serve as buffers against zoonotic diseases, which are illnesses transmitted between animals and humans. Deforestation disrupts these natural barriers, increasing the risk of disease transmission from wildlife to humans. The loss of forest cover has been linked to outbreaks of diseases such as Ebola and Zika virus.[10]

Deforesting our medical cabinet, and the home of the indigenous communities who hold much of this traditional plant knowledge, is shooting ourselves in the foot. Potential medicines are underneath the forest canopy, waiting for us to take notice. However, the commercialization of traditional medicines can also lead to overexploitation of natural resources in the region, as big pharma attempts to take a slice of the pie.

In the face of such existential battles, indigenous communities continue to demonstrate inspirational resilience. Threatened with dispossession and cultural elimination, these groups have been forced to find new and innovative ways to preserve their languages, spiritual practices and traditional knowledge, having to adapt and shrink into secret or underground settings. The perseverance of these cultures is a testament to the spirit of indigenous determination to keep the wisdom of the forest alive.

It is a story of survival against adversity that calls for acknowledgment, restitution and solidarity with indigenous peoples from across the continents, as they continue to assert their rights. Indigenous leaders and activists elevate the voices of their ancestors and communities in their quest to reclaim their fundamental right of custody over their lands and the space to nurture their cultures.

The goal of indigenous equity both legally and culturally has a place as an intrinsic pillar in the fight against the numerous environmental and cultural crises facing global society today. Since the dawn of modern science, indigenous knowledge has often been beaten down into second place, but it is time to realize that new is not always best. There is a lot we can learn from those who have not forgotten the old ways. The time is now, as we stand at this critical juncture in human history, to use every tool in our box to save Earth's forests, and in doing so, save ourselves.

Chapter 12

Leaves of Change

———

If the doors of perception were cleansed, every thing would appear to man as it is: Infinite.

> – William Blake, *The Marriage of Heaven and Hell*

During a lockdown of the bubonic plague epidemic of 1665,[1] 23-year-old Isaac Newton retreated from his studies at Cambridge University to his family home. During this enforced hiatus spent around the undulating hills of rural Lincolnshire, Isaac underwent an intellectual metamorphosis that would change the trajectory of human progression.

Legend has it that while sitting at the foot of one of the trees in his family's orchard, he watched an apple drop from one of the branches that hung above him. What followed this commonplace sight was a question that others before him had failed to ask – why did the object fall to the ground, and not sideways or even up towards the sky?

This question went on to produce one of the most profound bodies of scientific discovery in history. Upon his return to Cambridge, he would go on to make numerous contributions to physics and mathematics that still stand today, including the three Laws of

Motion, instigating the concept of white light as a combination of all wavelengths, and developing calculus techniques. But it was that innocuous day beneath the apple tree that led to Newton deciphering one of Earth's fundamental truths: the theory of gravity.

Newton's theory explains the mechanics of existence. He began to formulate connections between the invisible force grounding life on Earth and those holding the planet in a continual orbit with cosmic objects like the sun. Since its publication in Newton's *Principia* in 1687, his theory of gravity stands as a foundational influence in the inception of modern understanding of the physical world and lays the groundwork for feats of humanity like space travel and satellite communication.

The unassuming apple tree that Newton sat beneath on that day has since become known as 'The Gravity Tree'. Humankind remains indebted to that tree as the humble origins of a brilliant mind, and as a reminder that the natural world is at the core of humanity's knowledge.

Newton was a rationalist, who believed in the existence of an all-powerful divine creator. He rejected the squabblings of religion, saying, 'Tis the temper of the hot and superstitious part of mankind,' and instead saw in the divine a phenomenon to be explored and understood. He had no time for awe or a fascination with the mysterious. For Newton, God and nature were one and the same.

The origin of most religions is to revere nature, not man. The shift from nature-centred spirituality to anthropocentric religions is a relatively recent phenomenon in human history. Nature is where all of humanity's greatest ideas and knowledge – like our understanding of gravity – originally came from.

Life on Earth as we know it exists because of trees, and yet for millennia we have been destroying the very thing that gives us the foundations of our existence. On my travels in the Amazon rainforest, a shaman once told me, 'Whatever the question, the answer lies in nature.' But in order to recognize this wisdom, we must start to cultivate a genuine respect for both the power and potential of the natural world to alleviate our suffering and help solve our man-made problems. It starts with letting go of fear and control.

Reframing the perspective of our place within nature as a collaborative force, not a domineering one, is a crucial step in creating a harmonious and flourishing future. I believe this is already beginning to happen. Now, we are looking back to nature not only for ideas, but for real solutions to global crises. This concept of Nature-Based Solutions (NBS),[2] which include strategies such as rewilding, is gaining traction around the world, with many pioneers in the UK. To combat climate change and protect our forests, governments and organizations must embrace this. By doing so, they can unlock transformative possibilities for our planet's future. So what is already being done?

Less than 50 miles from the bustle of London, the Knepp Castle Estate offers a striking example of rewilding in action. Eighteen years ago, this land was an expanse of arable fields. 'It was a biological desert,' said Isabella Tree, who inherited the estate. 'We were told there was no way we could farm it.' Today, thanks to Tree and Charlie Burrell's bold decision to abandon conventional farming in 2001, the estate has transformed into a thriving ecosystem. This ambitious venture has become one of Europe's most groundbreaking rewilding initiatives and the first large-scale rewilding project in lowland England. By removing 70 miles of internal fences and welcoming herds of English longhorn cattle and Tamworth pigs, they stepped back, letting nature take the lead.

What happened next was remarkable. Knepp now hosts the largest population of rare purple emperor butterflies in the UK and boasts 13 of the country's 17 bat species. Perhaps most astonishing is the resurgence of turtle doves, a species that has faced a 96 per cent decline since 1970, with 23 singing males spotted just last year. 'As soon as the conditions are right, these species will find you somehow,' said Tree. 'It was like a miracle.'

This rewilding project not only revitalizes the landscape but also inspires hope, showing that with intentional changes, we can restore biodiversity and our connection with nature in a surprisingly short period of time. In the face of escalating environmental challenges, there is a growing recognition that nature itself holds some of the most potent resolutions. All we have to do is give nature back the reins and empower her.

The solutions for helping to combat our environmental crises are, in essence, to return those ecosystems that can no longer support the life they once did to their former strength. Nature-based solutions leverage the inherent power of ecosystems to heal, mitigate climate change and restore balance, all while integrating ideas to address societal challenges and ensure surrounding communities are uplifted.

I wrote earlier about some of the environmental devastation I have witnessed on my travels – scorched lands and endless rows of palm oil plantations. But not all is lost. Amid this destruction, I have been inspired by many incredible beacons of hope.

Back in Borneo, my guide Dean and I left the palm oil plantations and ventured deep into the rainforest of Bukit Piton, where the WWF had initiated a forest restoration initiative. Over 12 years, they planted nearly 350,000 native trees, including pioneer species resilient to harsh sunlight and poor soils, as well as fruit trees to support orangutans and other species like hornbills, aiding in natural regeneration.

Where years of destruction had left the landscape barren, now a lush ecosystem thrived. The project introduced 55 indigenous tree species, which is almost as many as the total native tree and shrub species in Great Britain. To date, 1,195 tree species have been documented in the landscape. The transformation was stunning to witness. Today, Bukit Piton stands as a restored forest.

But this is only part of the larger effort underway in Sabah. Just north of Bukit Piton lies the Malua Forest Reserve, where a ten-year forest rehabilitation project, from 2008 to 2018, successfully restored 5,400 hectares of degraded land. Over 300,000 seedlings from more than 95[3] native species were planted, many of them essential pillar species – those that help build a forest's resilience against fire and other threats.

This restoration has not just brought back the trees. It has had a profound impact on wildlife too. In Bukit Piton, my guide Dean took me in search of the many species this reserve has to offer: elephants, wild pigs and orangutans. As we walked, he read the forest floor for signs of the magnificent animals. 'I have been worried for a long time that we might lose them forever, but now I am convinced that positive change can happen, if we give nature time to heal,' said Dean.

As we continued our journey through Borneo, we arrived at a place that felt like the pinnacle of everything I had seen in restored forests; a glimpse of what these ecosystems once were, and what they strive to be again. One of the incredible examples of primary, untouched rainforest in the world, the Danum Valley is like something out of

Jurassic Park. It is the largest lowland rainforest in Malaysian Borneo. It is almost three times older than the Amazon rainforest, and revered as one of the world's most complex ecosystems. This was the benchmark, the vision that every restoration effort should aim for.

Borneo is extraordinarily rich in biodiversity. It is home to 3,000 species of trees, 15,000 species of flowering plants, 221 species of terrestrial mammals (including 13 species of primate), and more than 400 species of birds. Here, you can really feel it. Dean had heard rumours of a rare, dominant male orangutan in search of a breeding partner. 'If it's a flanged male then we're hitting the jackpot, because they cover a huge area up to five square kilometres, so they can be anywhere.'

Dominant flanged males are around seven times stronger than humans, with a bite force twice as powerful as a leopard. After hours of searching, we finally saw him. An enormous red man of the jungle peering down at us from the canopy. This encounter made the whole journey worthwhile. This is where wildlife truly thrives. While we continue to restore degraded environments and return them to nature, the best nature-based solution of all is to protect the last fragments of primary rainforests we have left, at all costs.

An orangutan was not the only thing I was excited to see here. The world's tallest tropical tree, at a whopping 100.8 metres, was discovered in 2019 in the Danum Valley. If it were laid to the ground, it would be the size of a football field! The team of scientists who found it dubbed the tree 'Menara', which in Malaysian means 'tower'. The newly discovered tree is nestled in a small, steep hollow. Scientists attribute its record height to the damp soils within the hollow and wind protection from a nearby ridge. Based on current theories regarding wind stress tolerance and hydraulic limitations in trees, scientists speculate that Menara may be nearing the maximum potential height for an Angiosperm anywhere on Earth.

The Danum Valley is home to many of these towering, record-breaking trees, all the same species – yellow meranti. This species is critically endangered and listed on the IUCN Red List due to relentless harvesting over decades. In the Danum Valley, they are protected, but yellow meranti logging persists in other parts of Borneo, often for concrete moulds and cheap plywood. Each of these remarkable trees is a biodiversity hotspot, hosting up to one thousand insect, fungi and plant species. Human intervention through nature-based solutions, be that preserving environments or restoring them, is the answer to saving these amazing trees. While nature does the hard work, it does sometimes need a helping hand from us.

A recent study, conducted in Sabah, used satellite imagery to track forest recovery at a 500-hectare experimental site, logged in the 1980s and designed to test the effects of replanting logged forests with native species. The site was divided into sections, and 20 years later, previously logged and replanted sites saw accelerated recovery when compared to those without human intervention and artificial seeding of native species. Sites where researchers planted a diverse mix of native species showed significantly faster recovery in canopy size and overall biomass than those without intervention. Even those sections with just one tree species planted recovered faster than those left to regenerate naturally.[4]

Oxford University's Professor Andrew Hector affirms that choosing to replant with a diverse array of native species has multiple ecosystem wins, including boosting biodiversity and carbon sequestration. The findings underscore how humanity is capable of positively impacting those sites we have wrecked, and emphasize the necessity to conserve undisturbed biodiversity to aid in future restoration efforts. After all, tropical forests are home to more than 50 per cent[5] of all biodiversity on Earth.

To make these initiatives happen, people have come up with creative ways to motivate large organizations financially to invest in them and make them successful. In the Democratic Republic of the Congo, Tellus Conservation are doing just that. Co-founded by former British army officers Jody Bragger and Angus Aitken, their project in the DRC works with local NGO Foundation Bombo Lumene and harnesses the power of carbon credits to restore a sprawling 2,800-square-kilometre national park.

Bombo Lumene, in the west of the DRC, is a complex environment of dense tropical forest and open savannah. The park had lost 77 per cent of its tree cover over the last 20 years, predominantly for cooking charcoal. The scale of this destruction meant that this once-thriving ecosystem faced an uncertain future. Tellus Conservation, recognizing the potential of nature-based carbon credits as a tool for both environmental and economic rejuvenation, has put together a plan that is set to become one of Africa's great conservation stories.

Surrounded by towering giants of trees in one of the last remaining swathes of forest in the park, Jody Bragger cuts a dashing figure. Even with the buzz of chainsaws in the background, the insatiable demand for charcoal in the markets of Kinshasa being unrelenting, he was overwhelmingly positive about the future.

Jody told me that Bombo Lumene represented a paradigm shift in conservation finance, showing how carbon credits can be leveraged to achieve both environmental and socioeconomic objectives. They hoped to set a precedent for sustainable development and environmental stewardship and prove that if they could do a project like this in the Congo, returning the forest to its former glory while still making a return for an investor, it could be done anywhere.

Tellus Conservation use precise methods to measure carbon stored in the trees and soil, establishing a baseline for how much carbon the park absorbs. By reforesting degraded areas and employing sustainable land management, they increase the park's capacity to capture carbon from the atmosphere. This process earns carbon credits, essentially permits that represent a reduction of one ton of carbon dioxide emissions. These credits are valuable, because companies and countries looking to offset their own emissions can buy them on the global market. This system provides vital funding for conservation by turning the carbon sequestered by trees into a financial asset.

But it is not solely about offsetting emissions. The revenue from selling these carbon credits supports further conservation work, while also benefiting local communities. By generating an income stream, Tellus Conservation creates real economic incentives for people living near the park to get involved in protecting it. The project offers alternative livelihoods, which strengthens the community's resilience against economic and environmental pressures.

Bombo Lumene is a powerful example of how nature-based solutions, like reforestation, can tackle multiple environmental issues simultaneously. Beyond storing carbon, the restoration efforts revive wildlife habitats, prevent soil erosion, regulate water cycles and protect biodiversity. It is proof that safeguarding the environment and supporting communities can go hand in hand.

The funds from purchasing carbon credits go not only to nature-based solutions like Bombo Lumene, but also to projects such as renewable energy development, reforestation or conservation initiatives, which are verified for authenticity through standards like the Verified Carbon Standard (VCS) programme by Verra, the largest certifier of these credits. Once a credit is purchased, it is permanently 'retired', meaning it can no longer be used, ensuring that its environmental benefit is realized.

This works incredibly well when projects are spearheaded by people like Jody and Angus, who are genuinely passionate about protecting the environment and local communities. But while the concept of carbon credits may seem like a fair solution – that is not always the case.

New research declares the damning reality of these carbon pay-offs.[6][7] According to a study published in *Science* in 2023, of the 89 million deforestation avoidance projects that were sold as carbon credits, and providing 40 per cent of Verra's verified carbon credits, only 5.4 million or 6 per cent had actual links to carbon reductions through tree conservation.

Researchers claim the reasons for these vast misjudgements come from opportunistic inflation, attempts to maximize revenues from credits and corruption. According to Professor Andreas Kontoleon of Cambridge University, 'there are perverse incentives to generate huge numbers of carbon credits, and at the moment the market is essentially unregulated.'[8] It seems that until the industry works to close its loopholes, it leaves room for those of bad faith to operate in an exploitative manner.

The only way this system can work and help to save and preserve forests is if more diligent organizations like Tellus Conservation are championed. While these economic models are a step in the right direction to prioritize trees and natural ecosystems in addressing climate change, their effectiveness depends on proper implementation and accountability. Yet, they still give us hope that nature-based solutions are being championed across the world in innovative ways.

But nature-based solutions are not simply about restoring forests. Reintroducing trees to areas that have experienced severe degradation and desertification can also be transformative. Across the globe, restoration projects have attempted to breathe life back into once-barren landscapes. The Great Green Wall initiative aims to halt desertification along the southern edge of the Sahara by planting trees and shrubs to protect and enrich the land. Originally set to span 7,775 kilometres (4,831 miles) from Djibouti to Dakar, the initiative has expanded to include additional countries in northern and western Africa.

However, such projects are not without their challenges. The dangers of relying on non-native monocultures were starkly illustrated in 2000, when a billion poplar trees in China's Ningxia region fell victim to disease in a single season, wiping out two decades of effort and highlighting the importance of biodiversity in restoration. These efforts are most successful when biodiversity is prioritized over monocultures.

A great example of successful restoration comes from China's Loess Plateau, where a grazing ban allowed native vegetation to double in distribution. This transformation turned severely eroded land into a productive, resilient landscape. When nature is given the space and time to recover, it can reclaim even the most degraded environments.[9] But sometimes even more of a human prod is required. In Saudi Arabia, environmentalists set out to try and fight desertification in remarkable ways.

But how do you create a forest in a desolate landscape with no water and no soil? It is a bleak starting point. The Al Baydha Project, in rural, western Saudi Arabia, is a land restoration, poverty alleviation and heritage preservation programme. The initial challenges were daunting: the area experienced minimal rainfall, with only four significant rain events in the first three years. However, the team responded with ingenuity, developing a series of low-tech channels designed to capture the scarce rainfall.

This approach not only directed water to where it was needed most, but also helped restore aquifers. After nine years of dedicated effort, the transformation is striking. The once-barren terrain has evolved into a savannah-like environment, flourishing with grasses and small trees.

Seeing the before and after pictures of this process of bringing back life to the region is incredibly moving. 'Nature is coming back, our mineral and water cycles are healing, and an upward spiral of life compounding on life has replaced the downward spiral of desertification,' said co-founder of the project, Neal Spackman. The project is an inspiring example of how human ingenuity and thinking beyond conventional management techniques to work with nature can lead to ecological restoration that was not thought possible before.

Across the Red Sea in Egypt, a Dutch firm of 'holistic engineers' known as the Weathermakers are literally trying to use plants to change the weather in the Sinai Peninsula. They aim to restore the Sinai Peninsula's ecosystem by reviving its natural water cycle. Their project starts with Bardawil Lake, where they are dredging sediment to improve water quality and fish populations. This initial effort is designed to jumpstart a larger goal: regreening the desert by using the dredged materials to plant vegetation.

Their theory is that by reintroducing plants, the landscape can retain moisture, which will restore rainfall and potentially shift weather patterns across the region. They argue that restoring natural systems can reverse desertification and mitigate climate change – showing that working with nature at scale, instead of relying only on technological fixes, offers a profound solution to environmental challenges.

We are fortunate to live in a world with some incredibly creative minds that choose to dedicate their skills to bolstering the natural world and fighting against the environmental crises that threaten human life on Earth. Technological evolution is becoming increasingly valuable to nature-based conservation efforts, both in the protection of existing habitats and in helping to restore others.

The pressing need for conservation, restoration and adaptation strategies has paved the way for some of the largest leaps in conservation science seen in the 21st century – one of which being the development of unmanned drone technology. Drones and satellites are providing researchers with the ability to gather vital and previously unattainable information, such as the extent of illegal logging activities in old-growth forests and overall forest health, in those locations that need it the most.

One region of great interest is the Great Western Woodlands in the south-west of Australia. It is the largest temperate woodland ecosystem on Earth, home to over 3,000 species of flowering plants, and boasts a truly unique global biodiversity hotspot. The woodlands have been a thing of mystery to researchers for years, but face some of the most severe impacts from climate change, including a voracious increase in wildfires.

Yet in 2023, scientists[10] were able to use information collected from unmanned drones, satellites and LiDAR (a system that uses pulsed lasers to create 3D representations of these ecosystems) to generate a full image for the first time of the health of this old-growth forest, as well as determine which areas were suffering with the most prolific wildfires. These findings will help them decide which regions of the woodland to prioritize for wildfire management, including indigenous cultural burning practices.

But drones are not just useful for analysing forests. Often, degraded forests around the world are unable to recover alone, but are so large that the logistics of rallying enough volunteers and getting to them would be a herculean task. More importantly, these schemes would take far too long to become effective in the timeframe necessary to address the environmental crises we are facing. So, in the absence of Hercules, it looks as though drones are stepping in.

Scientists, and more recently start-up companies, are generating autonomous drones that are able to transport and deposit native seeds along a predefined route – acting as a swarm to drop these seeds in remote, hazardous or simply huge areas primed for reforesting potential. While these methods have yet to gain traction in the tropics, they are popping up all over within industrialized countries. AirSeed Technologies,[11] an Australian start-up, have begun reforesting the colossal forests of New South Wales. AirSeed's drones, equipped with artificial intelligence to record the location of each deposited seed pod, drop up to 40,000 pods per day – 25 times faster than hand-planting seedlings.

The biggest hurdle that these innovations face is that, seed-for-seed, aerial seeding is less effective. However, due to the volume of seeds able to be dispersed, this is still a viable tactic. To tackle the problems with seeds reaching germination once they make landfall, it is an exercise in biomimicry; creating a pod system that replicates what happens in nature. This involves timing the drops correctly with seasonal changes, as well as generating seed pods that contain the perfect balance of nutrients, carbon and microbes, each tailored to the different species of seed. These methods aim to give the potential saplings the best start in life.

These methods are becoming increasingly accessible to NGOs and other wildlife organizations. In light of the 2019–20 bushfires of Australia that saw over 19 million hectares burned, WWF-Australia partnered with environmental tech firm Dendra Systems to reforest over 50,000 hectares[12] of degraded Australian forest using aerial drone-seeding technology, in hopes of helping to restore vital koala habitats.

Drones are not the only piece of airborne tech that is helping to solve our environmental crises. As of 2023, the European Space Agency announced their plans to launch a satellite, *Biomass*, out of French Guyana in 2025. This satellite differs from anything else currently available to scientists, and will provide the most comprehensive monitoring of global forests ever created – but what does it do?

Biomass will observe the overall health of our forests, delivering critical information on how they are changing in response to climate change, their maturity, and even the global mass of trees, producing 3D imaging of our planet's forests with a resolution of 200m. It will determine how much oxygen they are producing, and how much CO_2 they are absorbing, but also measure how carbon fluxes change with forest regrowth and restoration.

In time, this satellite will provide crucial information of the effectiveness of forest restoration projects, what factors make them more successful, and how we can alter restoration techniques to better stave off climate change.[13] Most crucially, this project will provide uninterrupted monitoring of our forests for eight years, allowing a clearer understanding of their growth rates under increasingly challenging climates.

Individual people can help with this monitoring too. Perhaps not quite on the scale of the *Biomass* satellite! But there is always a place for local action and technological innovation can amplify citizen science. Scientists that study trees, and other plant and animal species, benefit from data collected by nature enthusiasts. One particular application, iNaturalist, allows you to share your observations with scientists and researchers around the world.

Not only can they help you learn about the species around your local environment or wherever you happen to be travelling, but the data you collect is useful for biodiversity assessments, scientific research and conservation. The City of Toronto is using data collected through iNaturalist to identify old-growth trees in the city, from which they plan to harvest saplings. These trees will be well-adapted to Toronto's climate, and thus resilient trees for the city to plant. Citizens' science tree data is especially helpful in the face of changing climate conditions, as each year seems to bring unprecedented firsts and it is even hard for scientists to keep up. For instance, 'range migration' means that trees are moving at their own behest to more suitable climates.

Over the last 30 years in the US,[14] both hardwood and softwood trees have been observed in new patterns of migration. Hardwoods like red maple, scarlet oak and sweetbay magnolia are moving west by 1.5 kilometres per year; and softwoods like red pine, short-leaved pine and bald cypress are transitioning by growing one kilometre to the north each year. Trees are no longer rooted where we thought, they are explorers in their own right.

But while restoring, preserving and documenting forests is all well and good, the reality is that human beings still need timber as a resource. Sustainable forestry practices provide a solution by allowing for responsible wood extraction while safeguarding forest health. By balancing resource use with ecological conservation, sustainable management techniques – such as selective logging – enable forests to regenerate and support biodiversity. Countries such as Sweden are models for this approach. The Swedish Forestry Act mandates sustainable forest management, ensuring that logging operations do not compromise biodiversity or the long-term health of forest ecosystems.

In recent years, more research into forest ecosystem dynamics has revealed better ways to log. Peter Wohlleben, author of *The Hidden Life of Trees*, advocates for forestry practices that mimic natural processes, such as selective logging and maintaining mixed-species forests. He argues that clear-cutting and monoculture plantations harm ecosystems and biodiversity, while selective logging preserves habitat and allows forests to regenerate naturally. Respect for trees must be at the heart of this process. And old-growth forests must be left alone at all costs. Well-managed forests can act as carbon sinks, but the key is that the forest must be given the opportunity to regenerate properly and biodiversity must be sustained.

Understanding the dynamics of a forest ecosystem is vital if we are to create a world where we live in partnership with trees. Research from a study of around 700,000 trees worldwide suggests that the longer they are left alone, the faster they grow, and the more biomass they support. It is not enough to grow trees to a certain age before hacking them down and starting all over again. Remember that trees pass on their past experiences and knowledge to others. Older trees protect the growth of new trees. When we cut a forest down to stumps, we reset the ecology of that area to almost zero.

Consumers must advocate for responsibly sourced wood products and support initiatives that prioritize forest conservation and restoration, putting a stop to illegal logging, but also ensuring that commercial forestry is done with the wellbeing and longevity of forests at its heart.

The reality is that as long as humans exist on Earth, there will be a demand for wood, pulp, and other tree and forest products, which a myriad of businesses will attempt to meet. Creating legislation that encourages sustainable logging practices is realistically one of the only ways that conservation goals will be widely met. While they do not have a perfect record, organizations like the Rainforest Alliance and the FSC have been fine-tuning these methods since the 1980s, which ultimately hinge on balance.

From an ecological perspective, balance is the cornerstone of sustainable forestry. It is the extent to which these forestry practices mimic the natural pattern of disturbance and regrowth; balancing the needs of the ecosystem, such as wildlife, with sufficient resource extraction to support those communities that rely on the forest for income, all while working to conserve our forests for the future.

There are numerous corporations, worldwide organizations and governing bodies all playing their part to combat deforestation, proactively fighting to protect those remaining pristine environments and expand our green spaces. Every form of nature-oriented action is worthy of celebration; with commitments ranging from global reforestation projects that require herculean efforts and alliances, to local councils making small changes to their legislation, yet yielding impacts that traverse generations.

These local and regional efforts are part of a broader global movement, with initiatives such as the Trillion Tree Campaign and the New York Declaration on Forests setting ambitious targets to reverse deforestation.

The Trillion Tree Initiative is the brainchild of three conservation giants: the World Wildlife Fund (WWF), BirdLife International, and Wildlife Conservation Society. Founded on a vision of a planet in which our forest cover may expand, the Trillion Trees venture brings together the true masters of conservation mobilization through a network that funnels the right funds to the right places.

What I find interesting about the Trillion Tree project is its acceptance that in certain areas, economic development must go hand-in-hand with conservation to see long-term success. It can feel almost shameful to place a monetary value on these bastions of human life, but when deforestation and land degradation cost the world $6.3 trillion each year,[15] we have to accept the financial roles they play in our global system that can act as the leverage we need for prominent change.

They also place a focus on safeguarding the livelihoods of those who inhabit the surrounding regions of their projects to develop long-term, social and ecosystem-centric solutions. Whether this be establishing synergistic tea plantations within the 36 million trees the scheme protects deep in the Atlantic Forest of Brazil, Paraguay and Argentina, or developing a community-run cocoa harvesting business for the 140,000 indigenous Ngoleagorbu that thrive around the edges of the Gola forest in West Africa.[16]

As with many of these large-scale initiatives, it is not without its critics. 'Corporations are greenwashing us when they say they will achieve net zero, if that is relying on removing carbon through tree planting,' says Kate Dooley, a lecturer at Melbourne University.[17]

Climate scientist and director at California's Breakthrough Institute, Zeke Hausfather, also sees problems with the strategy. 'Reforestation might buy us up to a decade of time,' he said, 'maybe six or seven years of current emissions. It's not nothing, but it doesn't really fundamentally change the story: we still need some pretty massive reductions in our emissions. The brutal maths of climate change is that, as long as emissions are above zero, the world will continue to warm. You're going to max out pretty soon if you only try to use forests. Tree planting is not an alternative to mitigation.'

Despite this, tree planting has an incredibly important place in fighting the climate crisis; it just needs to be harnessed alongside other emission-reducing initiatives and not presented as the sole solution. A lot of this strategy comes in the form of regulation and government action.

Launched at the 2014 United Nations Climate Summit, the New York Declaration on Forests (NYDF) is a significant partnership that unites governments, corporations, indigenous groups and NGOs to combat deforestation. Signatories have committed to ambitious goals, including halting deforestation by 2030 and restoring 350 million hectares of forest.

This collaboration illustrates how diverse stakeholders can collectively advance global forest conservation, with efforts in reforestation aiming to create forest areas equivalent to the size of Peru across 36 countries. While there has been some progress, with global deforestation rates declining, continued and accelerated action is essential to meet the NYDF's targets. Recent studies indicate that significant reductions in deforestation rates are necessary to achieve these 2030 goals.[18]

Nature-based solutions are not confined to pristine wilderness areas; they have equal importance when attributed to urban environments. As Matthew Collins says in *The Tree Atlas*: 'Trees are the great leveller between the rural and urban experience.'

Urban forests may not be the thrumming superorganisms of the Amazon rainforest, but they act as an oasis for pollinating insects and other keystone wildlife species, as well as uplifting the spirits and mental wellbeing of those inhabiting the city. It is not just in existing forests that tree planting is being championed; city planning regulations are also recognizing the importance of green spaces.

In 2018, the British government unveiled *A Green Future*, a 25-year environmental plan aimed at transforming how government bodies interact with the natural world. Its mission: to reconnect an increasingly urban population with nature, restore biodiversity and tackle climate change. This marked a shift in focus for the Forestry Commission, whose original mandate of producing timber was expanded to include nature restoration.

By 2023, the commission rolled out *Thriving for the Future*, a bold five-year strategy that recognized the urgent need to move beyond reforesting post-war landscapes. Now, the focus is on addressing the twin crises of climate change and ecological degradation – an opportunity for the UK's rainforests that we cannot afford to miss. Ever-expanding cities also have a role to play.

London stands as a remarkable example of a 'tree city', with approximately 8.4 million trees covering around 21 per cent of its land area. This extensive greenery earned London recognition in 2019 as part of the *Tree Cities of the World* programme, acknowledging its commitment to urban forestry. The trees of London play a crucial role in improving air quality, removing an estimated 2,241 tonnes of air pollutants annually and helping to mitigate the urban heat island effect.

The Mayor of London's environment strategy aims to plant over 2 million trees by 2050. The inception of London's 'Urban Greening Factor' in 2021[19] saw strict codes put into place that prioritize the planting of flora into new buildings and their surrounding areas, alongside the protection of fauna such as bats, birds and insects. Using the measurement of 'biodiversity net gain', the legislation declines permission for construction plans that do not contribute towards raising the biodiversity of the city, helping to set a new standard for the anthropogenic presence within our largest city.

Concepts including nature-based, 'biophilic' architecture are contributing to an increasingly sustainable and thriving urban environment, by re-envisioning our urban landscapes to become intertwined with greenery and wild accents. These novel greening methods include green roofs and living walls, which support populations of moss and other vegetation, often flowering grasses to support pollinators; while also reducing energy consumption by elevating insulation properties and improving air quality.

One place that exemplifies the power of biophilic design is Singapore Changi Airport, which stands out as one of my favourite airports in the world. On my last visit, butterflies flitted about in lush gardens and I wandered through a maze of hedges, lost in its verdant twists and turns. The highlight, without a doubt, is the Rain Vortex – the tallest indoor waterfall in the world – plunging seven storeys within the Jewel complex. It is a place that blends nature and creativity with travel. But it is not just the airport.

Singapore is a pioneer in biophilic urbanization, featuring over 50 per cent green cover across the city. Its iconic Supertree Grove, part of Gardens by the Bay, includes 18 vertical gardens reaching up to 50 metres tall. These structures harness solar energy and collect rainwater, while a misting system cools the surrounding area. The integrated conservatories, like the Flower Dome and Cloud Forest, create unique microclimates that support diverse plant species. This innovative approach enhances the city's aesthetics, improves air quality, and promotes biodiversity, making Singapore a model for sustainable urban development.

The importance of trees in the city for public health has been understood at least since King Cyrus the Great planted his beautiful 'royal garden' in the metropolis of Pasargadae, the capital of the Persian Empire, 2,500 years ago. The paradise garden was filled with fruit trees, roses, lilies and jasmine, all irrigated by streams and pools. The remains of the limestone channels can still be seen to this day in modern Iran.

Since that time, all the great cities tried with varying degrees of success to incorporate nature into the urban environment. The Hanging Gardens of Babylon were so remarkable they secured a place in the Seven Wonders of the Ancient World. Modern cities also recognize the importance of creating space for nature to flourish. New York City has Central Park, Paris has the Bois de Boulogne and London has Hyde Park.

Beyond mere aesthetics, there are some more practical applications to bringing trees into cities: the cooling influence of trees extends into urban landscapes. Research conducted by Jonas Schwaab and his team at the ETH university Zurich revealed that areas covered by trees within cities experience significantly lower land-surface temperatures compared to treeless areas. Analysing data from 293 cities across Europe, they found temperature variances ranging from 8°C to 12°C in central Europe and from 0°C to 4°C in southern Europe.[20] As cities deal with rising temperatures due to global warming, these findings reveal the invaluable role of trees in mitigating urban heat islands.

Trees cool the air through evapotranspiration, and act as air purifiers for streets in cities. Next time you stroll through a sunlit park, pause to rest beneath a majestic old tree. As you recline in its shade, notice a refreshing temperature drop; this cool oasis is a result of the mature deciduous tree releasing a substantial amount of water from its leaves. This water, evaporating into the air, mimics the cooling effect of our body's sweat, creating a naturally refreshing ambiance.

In studies on urban heat island effect, researchers have examined cities around the globe, block by block, to understand heat disparities, or locations where heavy tree cover can be significantly cooler than concrete areas. In one study[21] looking at greenery in Hong Kong, shade provided by trees was able to reduce the temperature by 18°C compared to non-shaded areas.

Hong Kong is also a remarkable example of a high-density urban area that has been able to maintain considerable forest cover, such as

Victoria Park, that still makes up the central part of Hong Kong main island. Because of this tree cover, Hong Kong also consistently rises to the top of the list of urban areas with the highest levels of biodiversity.

People in cities like Hong Kong have access to nature, which has benefits that extend even to better population and lower crime rates. Nature calms us. We are more humane within an environment of trees. The oxygen fills our lungs and allows our nervous system to relax. Our brains and hearts are literally soothed with the rhythms of the forest. Research suggests[22] people are less violent when they live near trees.

One of the oft-cited examples is a study that looks at women in a Chicago housing estate. Those who lived near the trees reported less mental fatigue and less violent tendencies than those in barren areas of the same estate. In a clinical study on the impact of trees adjacent to hospitals, the researchers found that 'patients with bedside windows looking out on leafy trees healed, on average, a day faster, needed significantly less pain medication and had fewer postsurgical complications than patients who instead saw a brick wall.'[23] Maybe we should add another verse to that age-old adage: 'a tree-viewing a day makes the doctor go away'.

It might be stating the obvious, but a simple walk outside among the trees – just like our ancestors did – can save us from the over-taxation of modern life.

In Toronto, psychology professor Marc Berman used data sets to study the impact that the proximity of trees had on mental health. His work revealed that the presence of ten or more trees in a city block had the equivalent positive effect on a person's mental wellbeing as having an increase in salary of £10,000 or more, or the ability to feel seven years younger. So if you're feeling old, go and plant a few trees outside your window.[24] If that is not a strong enough argument for green city planning, I don't know what is.

A compelling study from the University of Glasgow demonstrated that trees reduce health inequality. 'Health inequalities related to income deprivation in all-cause mortality and mortality from circulatory diseases were lower in populations living in the greenest areas,' said Rich Mitchell, a public health professor at the university.[25]

In his work at the US Forest Service, researcher Geoffrey Donovan found correlation between deaths in the tree population and human populations. Areas with ash trees decimated by the emerald ash borer beetle were also seen to have increases in human mortality rates. Donovan refers to trees as 'a matter of life and death'. They can help us just as much as we can help them.

But being an advocate for trees extends beyond planting them in cities. As major corporations continue to expand their reach, the environmental impact of their operations often overshadows local efforts for sustainability. It is important that we bridge the gap between urban ecological advancements and the broader, often detrimental effects of global commerce on our planet.

Our world is run off the back of retail and industrial corporations, which hold the most influence over the way we use the Earth's resources and forge our consumer habits. So, it is no surprise that their changes in attitude and activity will have some of the greatest impact on our battle to fight deforestation and climate change.

From reconstructing supply chains to encouraging responsible consumer choices, the creation of alternative distribution methods at the hands of these corporations is helping reshape how we interact with forest resources and land. Certain companies are helping to spread education and awareness to their consumers, a mammoth task that is keeping the momentum to generate an inspired youth, the future environmental stewards who revere and understand the gravity of our forests.

The global supply chain is a complex web, connecting products from source to consumer across the world. It is this very system that holds most of the responsibility for the current state of our forests, contributing to deforestation and environmental degradation through resource extraction or logistical development (such as the building of roads and manufacturing plants). Traditional supply chains are often opaque, clouding the origin traceability of products. This lack of transparency allows for the inclusion of deforestation-linked commodities, such as palm oil, soy and beef, in various consumer goods.

However, innovative changes to distribution methods are emerging. Increasing the availability of data on these corporations through technologies like supply chain transparency, blockchain resource tracking and sustainable source certifications, is revolutionizing the way products are tracked and sourced. This empowers consumers to make informed choices on the products they purchase, enabling them to reduce the demand for those irresponsibly sourced products.

Richard Walker is the executive chairman of Iceland Foods, a British, family-run supermarket chain where my mum used to shop. He is also a bold advocate of corporate responsibility. We met in a coffee shop in the City of London to discuss his plans to fight climate change.

'Through time spent within the business I realized there were many environmental problems, from deforestation to carbon emissions – to which Iceland was not only a contributor, but also a potential changemaker.'

Walker, no ordinary greengrocer, is a pioneer in the field of corporate activism. He has climbed Mount Everest, supports Greenpeace, runs marathons and built an office block in London made entirely of sustainable timber.

'It occurred to me that as a business serving the most deprived communities around the UK, we have an authority – an obligation even – to advocate for environmental solutions that are both applicable and just, for everybody. We know what we need to do: reduce carbon and restore nature, alongside creating greater social equality.'

'These debates were had a decade ago, and I'm frustrated that they haven't really moved on. I suppose it's easier to talk platitudes, wring our hands and feel guilty, rather than focus on specific actions that would actually deliver it,' he says in his book *The Green Grocer*.[26] 'I'm sick of being made to feel shocked and horrified about what's going on in our world. Eco-alarmism doesn't help anyway – when you become overwhelmed by an issue you feel powerless. Despair and pessimism is easy. What we need now is leadership . . . it is about taking action.'

Thankfully Walker is a man of his word. He set up the Iceland Foods Charitable Foundation (IFCF)[27] to tackle the environmental crises within the industry, and his plans are ambitious. Backed by the likes of the Wildlife Trusts Wales and the National Trust, IFCF is restoring an ecosystem of incalculable worth, too often overlooked by conservation giants.

Peatlands store the most carbon of any other landscape, taking up only three per cent of global terrestrial land yet harbouring over thirty per cent of the world's carbon. Restoring peatlands locks carbon away, but also acts to filter terrestrial water stores, simultaneously tackling the UK water quality crisis. He has also cut out palm oil from most of his products in an effort to combat deforestation in Indonesia, for which he has received threats.

This sort of leadership does not only give back to a world from which it extracts, it cultivates a community of activists, fundraisers and campaigners who wish to enact change, providing them with a platform and the tools to inspire and foster that change. More companies should follow in Iceland's footsteps.

But it is not just governments and corporations who can make a difference. As individuals, we all have the power to become agents of the green revolution.

Chapter 13

Reforesting Our Minds

———

I believe in God. I just call it by another name: nature.
— Frank Lloyd Wright, during an interview in 1966

Seeing a tree being cut down is a pretty commonplace event. If you saw a local council worker with a chainsaw in a hi-vis vest on the street, you would probably think nothing of it – around 42 million trees are cut each day. Yet, on 28 September 2023, Britain awoke to the news of the destruction of one particular 150-year-old tree by two local men bearing an unfathomable grudge.

The Sycamore Gap tree rested in a dramatic dip of Hadrian's Wall, in the Northumberland National Park in the UK. A seemingly eternal being, cherished so dearly by all those that had made the pilgrimage to sit under its shade, it was suddenly gone. The outcry of grief surprised many, and yet the deep sense of loss rippled across the world.[1]

This tree had witnessed the pain of both World Wars, numerous love stories, proposals and blossoming friendships. It had served as a guardian to those crossing over, the chosen site for hundreds of ash scatterings and final goodbyes, and the catalyst for creative careers and writers; most famously, it sparked wonder for the English countryside in the 1991 blockbuster film *Robin Hood: Prince of Thieves.*

The widespread outrage and media attention following its destruction underscored a significant cultural shift: society is increasingly recognizing trees as a vital components of our environment, deserving of protection and care. This collective grief demonstrates that people truly do value the intrinsic worth of trees, and it is sad that we realize how much only when they are gone.

A little seed of resistance is beginning to take root in many of us. I often peer at the 'nature' shelves in bookshops, and over time I have seen them proliferating, overtaking the likes of travel and biographies; a pattern we are seeing across media, with the BBC's *Perfect Planet* (2021) attracting over 6 million viewers.[2] It gives me hope as people seek escapism from societal pressure by connecting once again to the natural world.

Yet this understanding needs to spread fast – and that is something we can all play a part in. It starts with the will of people and a trickle-up effect. Our collective actions and choices can directly impact those at the top, those who pull the strings and can sometimes feel deaf and out of reach. When enough people demand change, a tipping point is achieved, as defined by Malcolm Gladwell: 'We reframe the way we think about the world . . . The world . . . does not accord with our intuition. Those who are successful at creating social epidemics do not just do what they think is right. They deliberately test their intuitions.'

Gladwell goes on to say that behind the creation of tipping points is 'a bedrock belief that change is possible, that people can radically transform their behaviour or beliefs in the face of the right kind of impetus . . . Tipping Points are a reaffirmation of the potential for change and the power of intelligent action.'[3]

Humans are one collective force, living separate lives, with individual wants and desires, yet an innate, subconscious connection essential for our survival – like Pando. A quaking Aspen, Pando is a colonial organism that appears as 47,000 stems (or individual trees), but remains connected by a singular root system that spans 106 acres.[4] We share an important connection, both with each other and the earth beneath our feet.

I would argue that it is only in recent times that this significant philosophical change in thinking has taken root. In the past, we often viewed nature as a battlefield, where survival meant constant competition for resources. This idea of 'nature red in tooth and claw', a phrase popularized by Alfred, Lord Tennyson in his poem *In Memoriam A. H. H.* (1850), dominated much of our thinking. However, recent scientific discoveries have revealed a much more complex and cooperative reality. Ecosystems are not driven just by competition, but also by mutualism – where different species work together for mutual benefit.

This emerging understanding is leading to a global philosophical shift. Instead of seeing ourselves as separate from, or in opposition to, nature, we are beginning to recognize that we can thrive alongside it, as nature thrives together.

By acknowledging this interconnectedness and our kinship to trees, we take the first step to preventing their exploitation – and our downfall. They are at the soul of our ecology, their roots closely binding our planet together in a network of collective vitality. Rectifying the disconnect that modern society has made us feel with nature could be our synergistic salvation. Implementing an increasingly nature-oriented focus within our daily lives and policies is pivotal.

Spirituality and science do not have to be held so far apart. If we can open our minds to the merging of both, we would find many reassuring similarities. The significance of the interconnectedness of nature, which is now being proven by science, is old news to the indigenous communities who held this wisdom for generations.

In spirituality, the interconnectedness of all things in the universe is a concept widely shared, interpreted and presented in various forms. Thich Nhat Hanh, a Buddhist monk and peace activist, uses the term 'web of interbeing'; David Whyte, a poet and philosopher, coined the phrase 'conversational nature of reality', to describe a dialogue between all elements of the universe; and Robin Wall Kimmerer, a botanist and indigenous knowledge holder, says 'sacred reciprocity' to depict a respectful relationship between people and nature.

Think about it this way. Many of the great environmental challenges we face stem from our disconnection with nature in our modern lives. Whether consciously or not, we often prioritize short-term gains at the expense of the one thing that has the power to save or destroy us. If we remember that we are not alone, we are truly a part of a much greater whole, then we might have the courage to do what is right and in doing so, save ourselves.

We must start to notice nature again, to respect and cherish her, regardless of whatever else is going on in the world. Even in chaos, especially in chaos, it is important to remind yourself that nature should always be our priority, because without that respect, life is meaningless. We ignore nature at our peril.

Herodotus, a geographer of the ancient world in ~430 BCE, famously retold the story of Xerxes, King of Persia. While walking along the banks of the Meander on his journey to invade Greece, his gaze met a tree that abruptly startled him. Astonishingly, Xerxes halted his battle plans momentarily to take in the extraordinary beauty of the tree. Feeling so moved, and wishing to demonstrate his appreciation for nature, he adorned the tree with an abundance of gold and jewels, and even stationed one of his guards beneath its branches for protection.

Perhaps we can all heed this lesson. No matter whether you are invading Greece, working at the Central Bank, or teaching at a school, we can make time to appreciate and venerate nature. Looking forward, we must envisage a world where nature is central in decision-making, a world where future generations, like the Kernel taught his children, learn that growing and protecting trees is a part of everyday life.

Whether it is choosing to keep a tree in your garden that casts an irritating shade, or our governments prioritizing the health of our environment over economic development opportunities, retaining an appreciation for trees, like Xerxes, is a crucial everyday step to fight for a better world.

So, what can we all do?

Firstly we need to talk. We must have open, frank debate about the environmental subject matter. Climate change is not a partisan issue, it is a human one that affects each and every one of us. It is not a matter of left or right, green or blue or red, it is simply a conversation that we all need to be in on, because, as Sir David Attenborough said: 'We depend on the natural world for every mouthful of food we eat and every lungful of air we breathe. Our health depends on its health. We are now by far the most powerful single species that has ever existed on Earth. That power brings great responsibility . . . What humans do over the next fifty years will determine the fate of all life on the planet.'

As it stands people feel left behind, ignored, or overwhelmed. The world has plenty of problems – war, poverty, oppression, racism, sexism and economic inequality – surely that is enough to contend with?

The thing is, none of these problems are going to get any better if we keep thoughtlessly cutting down trees. They will probably get worse. And, like all of those issues, the environment crisis can be resolved with a very simple fix.

Kindness.

By being a bit more respectful, compassionate and understanding towards one another, we would solve the vast majority of the world's problems. By setting aside our own egos, prejudices and entrenched views, we might just stand a chance of dousing the flames of the world. And we already know that by embracing nature, being kind to the trees and reconnecting with the forests, we will possibly become better people in the process.

'There's a really simple thing that doesn't cost anything,' says John Deakin, head of trees at the National Trust, 'which is just notice the trees that are around you, the seedling in your garden, the tree that's in the park, or the one by the bus stop, and just take time to notice that it is there and what that does to your environment when you're beneath it. Remind yourself that it is cooling you in the sun or protecting you from a rain shower when you're walking your dog. That, for me, would be a great start.'

Someone who did just that is Merlin Hanbury-Tenison, author of *Our Oaken Bones*. After suffering from the trauma of his military service in Afghanistan, Merlin and his wife, Lizzie, moved from

London to Cabilla, his childhood farm, nestled in the heart of Bodmin Moor in Cornwall.

They soon faced unexpected challenges: a farm sinking deeper into debt and to their surprise, the realization that the overgrazed and damaged woods, which run through the valley and where Merlin had played in as a child, contained one of the UK's last remaining fragments of Atlantic temperate rainforest, a critically endangered ecosystem. Merlin felt a huge ecological responsibility to restore the beauty of the rainforest. They engaged their community in a monumental effort to inspire collective action and heal the forest and in doing so, they healed themselves.

'Imagine if nature was our medicine. If the outside world could repair our inside world, facilitating a great connection and love for ourselves – and others. This is why Cabilla started,' writes Merlin.[5] In response, they have transformed the land into a retreat space, where others can come to heal through nature too. The retreat brings together leading experts in yoga, breathwork, somatic therapy, sound healing and culinary arts, offering a sanctuary for those in need. It is a place where nature and people work together and protect and repair each other. While it is only a small pocket of the UK, stories like this are an inspiration, showing how things get better when we work together.

Cabilla started with the actions of just two people but brought together a whole community. Small choices can make a real difference. And it starts with understanding and accepting our role as the stewards of nature. We are here not to dominate but live in harmony.

When I spoke with the renowned environmental activist Julia Hill, who lived in the branches of her beloved Luna tree for two years, I asked her what kept her motivated to keep up the activism, when things always appear so bleak. Her words stuck with me: 'I am an ancestor of the future, so I better start living that way.'

But you do not need to live in the branches of a tree to take positive environmental action that will affect generations to come. One way to relate ourselves more deeply with the forests, especially for the city-dwelling majority, is to reflect on our interconnected world through basic food purchases. British environmentalist and futurist James Lovelock's Gaia theory describes Earth as a self-regulating, interconnected system where all living organisms and their environments work together to maintain balance.

This idea of interconnectedness can also apply to our global food system. The food we consume every day links us directly to the environment, often in ways we do not see. At any given time in your local grocery store, you can find apples from the USA, pears from China, bananas from Costa Rica, coffee from Colombia, and so forth. The food decisions we make, like eating peanut butter made with palm oil, or foods sprayed with heavy pesticides, have a direct 'butterfly effect' in the rainforests of Indonesia, and in the other locations where food is grown.

One of the challenges with modern lifestyles is that because we do not always see first-hand our impact on the forest, we lose sight of our interrelationship. Global and local supply chains that we support either directly (through local purchases), or indirectly (in international purchases), undeniably impact the planet's sustainability and wellbeing.

Feeling empowered in a world of 8 billion people can feel daunting, even more so in the face of global adversity. Yet, no single action is meaningless. Capitalism can be criticized for a lot of the damage caused to our planet, but free markets also hold a powerful answer to turning the tables, so long as we are willing to vote with our wallets. The power of consumer patterns is very real, and individual action accumulates into wide-scale changes in demand. Advocating for ethical practices, through your consumer choices, helps to place pressure on organizations to stop creating products that damage our environment and ecosystems. This is the principle of how plant-based lifestyles can help to fight deforestation.

This dietary and lifestyle commitment – abstinence from animal products such as milk and meat, as well as the use of leather products – has become increasingly popular in recent years, despite much conservative opposition. Whatever you think of veganism as a movement, its potential to curb land-use change and deforestation is hard to ignore.

By thinking more carefully about what we eat, we are also becoming more aware of nature and her impact on our lives. We should understand where our food comes from. This is becoming more challenging as the global population is living in an increasingly urbanized environment, a world where the greys of concrete are gradually filling the vision of every city dweller. But I would argue that this also means the perception of trees and their role in daily life takes on a newly realized, profound importance. Because of our urban living situations, we need trees more than ever.

Blending our urban spaces with the natural world transforms not only our environments, but also the way we feel and think. By planting a tree in your garden, on your balcony or in a local community space, and creating wildflower beds, cities are becoming vibrant oases that alleviate stress and mental fatigue. We can also encourage bees, who are responsible for pollinating around 75 per cent of the crops that produce fruits, nuts and seeds consumed by humans. Without them, ecosystems and food supplies would suffer, leading to significant impacts on global agriculture.

In cities, urban beekeeping and pollinator-friendly spaces like urban gardens and green roofs can help create safe havens for bees, while reducing air pollution that also affects their health. Their survival is directly linked to the health of our trees, food sources, and even our mental wellbeing, through the restorative power of nature. Natural elements in cities enhance our innate ability to navigate and connect with our surroundings, but the real magic lies in their power to spark social interactions.

It is these connections that breathe life into communities and make cities thrive. Individual actions can spark movements. As we evolve this integration of nature into our cities, it is clear that it is no longer just a luxury – it is an essential ingredient for the wellbeing of our minds and the health of our planet.

Become an advocate for trees in your city or town, plant your own, but also encourage your local council to plant and protect them too. Prioritize getting out of the busyness of the city when you can and take time to pause in nature. Forest bathing can become a part of daily life. Connecting with nature is like building an exercise habit – it strengthens both body and mind. It takes practice at first, but once it is routine, it naturally improves life. Be aware of the nature around you.

Our role as tree stewards is to listen to the patterns indicating the health of trees, and the interrelationships they support, both within and outside of cities. In *The Overstory*, Richard Powers writes: 'Life will cook; the seas will rise. The planet's lungs will be ripped out. And the law will let this happen, because harm was never imminent enough. Imminent, at the speed of people, is too late. The law must judge the imminent speed of trees.'

Think about how important nature is for our childhood. For me, at least, treehouses are places of imagination and creation. The act of climbing a tree, reaching and holding onto branches, brings us in intimate proximity to the smells of wood, moss and resident insects. No matter how advanced virtual technology is, I do not think there will ever be a replacement that could provide as many physical, physiological and emotional benefits as running and playing outside among a grove of trees.

Reconnection to nature will invariably come if it is taught properly in the classroom. If we explained to children the importance of trees – that they shaped our world, give us the air we breathe, create rain, seed the oceans and provide the foundations of our medicine, traditions and culture – then it might make it a little more difficult to cut them down.

It has been shown that children who spend time in green spaces between the ages of seven and twelve tend to think of nature as magical. It is these children who will grow up to become the adults most concerned about a lack of protection for nature. And for those children who did not spend time playing in the woods, well, they are far more likely to regard nature as hostile or irrelevant and are indifferent to its loss. By purging nature from children's lives, conservationist Isabella Tree reminds us, we are 'depriving the environment of its champions for the future'.[6]

Richard Louv, author of *Last Child in the Woods*, describes the impact on children who do not grow up surrounded by nature as 'nature deficit disorder', where a lack of time outdoors leads to behavioural issues and a diminished understanding of the environment. Without regular contact with nature, kids grow up caring less about it.

One of the biggest proponents of encouraging a nature-based education is the British monarch. As the Prince of Wales, he was a huge champion of the environment and in his 2010 book *Harmony*, the now King advocates for children spending more time outside and learning about the value of the great outdoors for our physical and emotional wellbeing. He is a supporter of holistic medicine, green architecture and rediscovering abandoned knowledge, and as he states in no uncertain terms: 'Can we afford to ignore nature?' We definitely cannot.

Last winter, I ventured back to my roots in Stoke-on-Trent. The ashen clouds of December loomed in the sky threatening that Midlands drizzle over the rooftops of the red-brick houses, stood like toy soldiers – uniform, forever unchanging. I had been invited by the University of Staffordshire to visit their newest educational facility, the Woodlands Forest School.

Amid an industrial wasteland, there was hope: a rewilded forest, marsh and grassland, and then in the middle of it all, a playground filled with excited children. The school itself looked like a futuristic Scandinavian outward-bound centre, built from sustainable timber and fuelled by the sun. Except today it was raining.

Completely oblivious to the miserable weather, the children in their yellow rubber wellies were itching with excitement. 'Rain is fun!' a little boy screeched to me, as I walked past him to greet Dr Jane Robb, a lecturer in outdoor learning and a biodiversity expert. As I peered through the door into an orderly classroom, I could make out another group enjoying their scheduled inside time, crafting creatures out of clay. One boy appeared to be fashioning what looked like a hedgehog out of a small ball of brown mud, carefully placing cut-off sticks in its back to create spines.

I heard laughter and whipped around to see a young girl careening across the courtyard on a tricycle made entirely out of wood. 'There's no plastic here,' Jane told me. The group in their overalls were desperate to join them in their playing, but that day, they had bigger tasks at hand.

'Today, we're going to plant an apple tree,' announced Jane.

Leading the charge, Jane marched ahead, the row of little children in giant overalls following her like ducklings into the wood-chipped play area, where I could make out some potted saplings ready to go into the ground.

I felt a small tug at my coat sleeve. Looking down I was met with a pair of very excited eyes.

'I put these carrots in here!' a little boy proclaimed proudly, pointing and dragging me towards a nearby vegetable patch.

Jane clapped her hands, silencing the group. I was charged with lifting the first sapling into its freshly dug hole, with all hands on deck to secure it into the wet soil with gentle pats – everyone must grab a handful. The dozen tiny hands made light work of it, and soon half a dozen more saplings were safely secured in their new homes.

The children finished their time in the woodland with a brief roll around in the mulch, where they were encouraged to 'look up at the sky, and listen for all the different birds', before being released for free 'playtime' to let their imaginations run amok.

Forest schools are gaining traction across the UK, as an ever-increasing number of parents want to raise their children with an education that champions a close relationship with the natural world.

It is a bond I feel lucky to have had nurtured by my own parents in a time when playing outside was still encouraged.

It is poignant to witness this connection with nature being cultivated in children so young, despite the challenges of living in a city. I only hope future generations get to enjoy going for a walk in the woods as much as I did. After saying farewell to the children, I went back to my parents' house to say goodbye before heading back to London. As I always did before leaving, I stepped onto the lawn and reached out to pat the reassuringly rough bark of the horse chestnut tree I had planted all those years ago. She was 32 this year, growing strong, and had started to bear fruit. I knew that whatever else was going on in the world, that tree would always remind me of home, of my family and my childhood spent walking in the woods. It was only one tree, but it symbolised a very human existence. I wish I had planted more.

The natural world is not a separate entity, but an interconnected web of which we are very much a part. In ancient times, the words door, tree and truth all had the same linguistic beginnings: *Deru*. A promise.

Remember the wisdom of Benki and the Amazon rainforest: 'We are the trees and they are us.' Their story is our story. It would seem as though we have known deep down for a very long time that the forests are indeed the doors to the truth, and in the trees themselves we will find the promise of a shared commitment to look after this planet.

Together, we can live happily ever after.

Acknowledgements

As a tree relies upon the energy of the sun, the nourishment of rain and the company of its kin to grow and flourish, this book would not have been possible without the assistance of many supportive hands and the energy of a multitude of creative minds.

I am grateful to all those who kindly gave their time in the creation of this book; to the countless incredible conservationists, scientists, academics and field workers who granted me interviews and whose published works provided the basis of my research. Thanks too to Benki Piyãko and the community at the Yorenka Tasorentsi Institute for inspiring me to begin this journey in the first place and to the conversations I had with Robert Street, Ryan Whitewolf, Mike Bysiek and Ryan Locati in the Amazon rainforest.

I am deeply grateful to Evangeline Modell for the innumerable hours she put in helping to research and edit, to Jessica Minocha for commissioning the book, as well as Rimsha Falak, Ailie Springall and Charlotte Sanders at Octopus books; to Barry Johnston for his detailed eye and to my agent Jo Cantello for her unwavering support over the years.

ACKNOWLEDGEMENTS

Thanks to Geraint Jones and Philip Booth for giving me advice on structure, Charles McBryde for his insights into mythology, to Zan Campbell, Katie Conlon, Danni Pollock, Samuel Nicholson, Alice Payne and Katherine Bayford for their help in the assembly of research and detailed notes.

And in no particular order I also extend my gratitude to Rob Croucher and Austin Raywood at the Explorers Club, Jo Dyson at the National Trust, Richard Deverell at Kew Gardens, Nell Jones at the Chelsea Physic Garden, Nigel Winser at the Royal Geographical Society, Maxime Friess and Olivia Gauss at Emerald Stays, Dr Jane Robb at the University of Staffordshire, Rudy Randa at the Boa Foundation, and Vivien Vilela and all the team at Aniwa; Bern Teo in Bali, Professor Iván Batún at the Universidad de Oriente in Yucatán. Professor Charles Watkins at the University of Nottingham, Jody Bragger, Angus Aitkin, Bayard Baron, Niki Page, Karen Murray, Edward Cadogan (9th Earl Cadogan), Stephanie Comer, Charlie Roberts and Richard Walker. Special thanks to Dave Luke and Neil Bonner, who have accompanied me on many forest adventures, and my brother Pete with whom I climbed several trees as a child.

And of course, thanks to my parents who planted the seeds of hope for a greener future in my mind many years ago.

Picture Credits

Methuselah Tree, California (**possibly**) (page 1 above): Levison
Wood

A Depiction of Calamophyton, Early Trees (page 1 below): Peter
Giesen

Sugi Tree Circles, Japan (page 2 above): Katsumi Tanaka/
Associated Press/Alamy Stock Photo

Crown Shyness (page 2 centre): Sergei Kornilev/Shutterstock

Australopithecus Afarensis in the Forest (page 2 below):
Historic Collection/Alamy Stock Photo

Adam's Tree, Iraq (page 3 above left): Levison Wood

Yggdrasil, Norse Mythology (page 3 above right): Norman B.
Leventhal Map Center (CC by 2.0)/Flickr

Green Man, Rosslyn Chapel (page 3 below): Photofires/
Dreamstime.com

Christ Depicted as the True Vine, Rosslyn Chapel (page 4 above):
Levison Wood

Moai, Easter Island (page 4 centre): Gábor Kovács/Dreamstime.com

The Author at Yaxha Pyramids, Guatemala (page 4 below): Simon
Buxton

HMS Victory, Battle of Trafalgar, 1805 (page 5 above): IanDagnall Computing/Alamy Stock Photo

Amazon Deforestation (page 5 centre): Carl De Souza/AFP via Getty Images

Hanging Gardens of Babylon (page 5 below): North Wind Picture Archives/Alamy Stock Photo

John Muir and Theodore Roosevelt, Mariposa Grove, 1903 (page 6 above): Library of Congress Prints and Photographs Division Washington, D.C.

Activist Julia Hill and Luna (page 6 centre): Yann Gamblin/Paris Match via Getty Images

Lahaina Banyan Tree, Hawaii (page 6 below): Levison Wood

Benki Piyãko, Acre, Brazil (page 7): Levison Wood

Sycamore Gap Tree, UK (page 8 above): Phillip Maguire/ Dreamstime.com

The Author, Staffordshire, UK (page 8 below): Levison Wood

Bibliography

1 **The Overstory** by Richard Powers

2 **Underland: A Deep Time Journey** by Robert Macfarlane

3 **Finding the Mother Tree: Discovering the Wisdom of the Forest** by Suzanne Simard

4 **The Treeline: The Last Forest and the Future of Life on Earth** by Ben Rawlence

5 **The Hidden Life of Trees: What They Feel, How They Communicate** by Peter Wohlleben

6 **Entangled Life: How Fungi Make Our Worlds, Change Our Minds & Shape Our Futures** by Merlin Sheldrake

7 **The Spell of the Sensuous: Perception and Language in a More-Than-Human World** by David Abram

8 **The Songs of Trees: Stories from Nature's Great Connectors** by David George Haskell

9 **Sacred Nature: Restoring Our Ancient Bond with the Natural World** by Karen Armstrong

10 **The Sixth Extinction: An Unnatural History** by Elizabeth Kolbert

11 **Jungle: How Tropical Forests Shaped the World – and Us** by Patrick Roberts

12 **The Arbornaut: A Life Discovering the Eighth Continent in the Trees Above Us** by Meg Lowman

13 **The Mountains of California** by John Muir

14 **A Trillion Trees: Restoring Our Forests by Trusting in Nature** by Fred Pearce

15 **Silent Earth: Averting the Insect Apocalypse** by Dave Goulson

16 **Our Oaken Bones** by Merlin Hanbury-Tenison

17 **Wild Signs and Star Paths: The Keys to Our Lost Sixth Sense** by Tristan Gooley

18 **Ever Green: Saving Big Forests to Save the Planet** by John W. Reid and Thomas E. Lovejoy

19 **Tree: A Life Story** by David Suzuki and Wayne Grady

20 **Trees, Woods and Forests: A Social and Cultural History** by Charles Watkins

21 **Tree Wisdom: The Definitive Guidebook to the Myth, Folklore, and Healing Power of Trees** by Jacqueline Memory Patterson

22 **The Green Grocer: One Man's Manifesto for Corporate Action on Climate and Sustainability** by Richard Walker

23 **Into the Forest: How Trees Can Help You Find Health and Happiness** by Qing Li

24 **The Wisdom of Trees: A History of Trees** by Max Adams

25 **Intelligence in Nature: An Inquiry into Knowledge** by Jeremy Narby

26 **The New Sylva: A Discourse of Forest and Orchard Trees for the Twenty-First Century** by Gabriel Hemery and Sarah Simblet

27 **How Forests Think: Toward an Anthropology Beyond the Human** by Eduardo Kohn

28 **The Songlines** by Bruce Chatwin

29 **Wilding: The Return of Nature to a British Farm** by Isabella Tree

30 **Ishmael** by Daniel Quinn

Source Notes

————

CHAPTER 1: IN THE BEGINNING

1. David Attenborough, *Life on Earth*, William Collins, 2018, p. 80.
2. https://www.nhm.ac.uk/discover/news/2024/march/earliest-fossilised-forest-discovered-in-somerset.html#:~:text=Traces%20of%20an%20ancient%20forest,trees%20ever%20discovered%20in%20Britain.
3. https://www.theguardian.com/science/2024/mar/06/worlds-oldest-fossilised-trees-discovered-along-devon-and-somerset-coast#:~:text=The%20world's%20oldest%20fossilised%20trees,found%20in%20New%20York%20state.
4. https://www.theguardian.com/science/2024/mar/06/worlds-oldest-fossilised-trees-discovered-along-devon-and-somerset-coast#:~:text=The%20world's%20oldest%20fossilised%20trees,found%20in%20New%20York%20state.
5. https://www.cam.ac.uk/stories/earths-earliest-forest-somerset> Earth had reached a pivotal moment in which its entire surface would be changed – all because of fallen twigs.
6. Patrick Roberts, *Jungle: How Tropical Forests Shaped World History – and Us*, Viking, 2021, p. 18.

CHAPTER 2: TREESPIRACY

1. David Abrams, *The Spell of the Sensuous*, Vintage, 1997, p. 16.
2. https://www.nature.com/articles/s41598-020-76588-z?utm_medium=affiliate&utm_source=commission_junction&utm_campaign=CONR_

PF018_ECOM_GL_PHSS_ALWYS_DEEPLINK&utm_content=
textlink&utm_term=PID100090912&CJEVENT=e62a2374fb7711ee809f
e9750a18b8f9.

3. Jeremy Narby, *Intelligence In Nature: An Inquiry Into Knowledge*, Tarcherperigree, 2005.

4. Suzanne Simard, *Finding the Mother Tree: Uncovering the Wisdom and Intelligence of the Forest*, Allen Lane, 2021, p. 4.

5. Merlin Hanbury-Tenison, *Our Oaken Bones: Reviving a Family, a Farm and Britain's Ancient Rainforests*, Witness, 2025, p. 50.

6. https://www.spun.earth/networks/mycorrhizal-fungi#:~:text=A%20 single%20gram%20of%20soil,with%20much%20of%20its% 20structure.

7. Peter Wohlleben, *The Hidden Life of Trees*, Greystone, 2016, p. vii.

8. https://www.ncbi.nlm.nih.gov/pmc/articles/PMC2954164/.

9. Richard Powers, *The Overstory – A Novel*, W W Norton & Company, 2018, p. 424.

10. John Muir, *The Mountains of California*, New York: The Century Co, 1894, p. 250.

11. https://bloomscape.com/green-living/does-music-affect-plant-growth/.

12. https://journalofyoungscientist.usamv.ro/pdf/vol_V_2017/Art12.pdf.

13. D L Retallack, *The Sound of Music and Plants*, DeVorss and Co., USA, 1973, p. 27.

CHAPTER 3: MOTHER EARTH

1. https://pubs.geoscienceworld.org/gsa/geology/article-abstract/37/10/875/ 103834/Chemical-constitution-of-a-Permian-Triassic?redirectedFrom= fulltext.

2. https://www.sciencenews.org/article/more-plants-survived-world-greatest-mass-extinction.

3. https://www.sciencedaily.com/releases/2023/09/230912192432. htm#:~:text=Whilst%20the%20fossil%20record%20shows,of%20 these%20are%20flowering%20plants.

4. Fred Pearce, *A Trillion Trees: How We Can Reforest Our World*, Granta, 2021, p. 19.

5. Peter Wohlleben, *The Power of Trees: How Ancient Forests Can Save Us If We Let Them*, Greystone, 2023, p. 66.

6. John W Reid & Thomas E Lovejoy, *Ever Green: Saving Big Forests to Save the Planet*, W W Norton & Company, 2022, p. 6.
7. Attenborough, *Life on Earth*, p. 88.
8. Wohlleben, *Hidden Life of Trees*, p. 2.

CHAPTER 4: OUR FIRST HOME

1. Attenborough, *Life on Earth*, p. 329.
2. Ben Rawlence, *The Treeline: The Last Forest and the Future of Life on Earth*, Jonathan Cape, 2022, p. 282.
3. Roberts, *Jungle*, p. 95.
4. https://www.nbcnews.com/id/wbna41319336.
5. https://www.cell.com/trends/ecology-evolution/fulltext/S0169-5347(18)30117-4?&.
6. https://the-past.com/feature/rethinking-the-jungle-the-forgotten-story-of-humanity-and-tropical-forests/#:~:text=While%20you%20eventually%20get%20some,first%20arrival%20in%20Sri%20Lanka.
7. Roberts, *Jungle*, p. 118.
8. Rawlence, *Treeline*, p. 282.
9. Rawlence, *Treeline*, p. 6.
10. https://comptes-rendus.academie-sciences.fr/geoscience/articles/10.1016/j.crte.2009.06.007/#article-div.

CHAPTER 5: THE ROOTS OF CIVILIZATION

1. https://www.nationofchange.org/2019/02/06/dominion-mistranslated/.
2. https://blogs.timesofisrael.com/on-judaism-and-the-environment/.
3. Karen Armstrong, *Sacred Nature*, Bodley Head, 2022, p. 119.
4. Charles Watkins, *Trees, Woods and Forests: A Social and Cultural History*, Reaktion, 2014, p. 31.
5. 2 King 17:10.
6. Watkins, *Trees, Woods and Forests*, p. 140.

CHAPTER 6: BROKEN BOUGHS

1. Wade Davis, *The Wayfinders*, House of Anansi Press, 2009, p. 85.
2. Chris D Thomas, *Inheritors of the Earth: How Nature Is Thriving in an Age of Extinction*, Allen Lane, 2017.

3. https://www.historytoday.com/archive/missing-pieces/lost-city.
4. https://www.newyorker.com/magazine/2005/09/19/the-lost-city-of-z.
5. https://www.newscientist.com/article/dn27945-myth-of-pristine-amazon-rainforest-busted-as-old-cities-reappear/.
6. Pearce, *A Trillion Trees*, p. 81.
7. https://www.ncbi.nlm.nih.gov/pmc/articles/PMC8785365.

CHAPTER 7: THE FOLLY OF MAN

1. https://diposit.ub.edu/dspace/bitstream/2445/101825/1/575274.pdf.
2. Lewis Dartnell, *Origins: How the Earth Shaped Human History*, Bodley Head, 2019, p. 256.
3. https://ecozagroza.gov.ua/.

CHAPTER 8: APOCALYPSE NOW

1. https://openknowledge.fao.org/server/api/core/bitstreams/b19b71f7-ed39-4dd9-98e6-5cfde0a8918e/content.
2. https://www.ipbes.net/news/Media-Release-Global-Assessment.
3. https://iucn.org/press-release/202410/more-one-three-tree-species-worldwide-faces-extinction-iucn-red-list.
4. https://www.cifor-icraf.org/publications/Corporate/FactSheet/fast_wood.htm.
5. https://news.mongabay.com/2023/04/mining-may-contribute-to-deforestation-more-than-previously-thought-report-says/.
6. https://www.pnas.org/doi/10.1073/pnas.1500415112.
7. Rawlence, *Treeline*, p. 282.
8. Rawlence, *Treeline*, p. 283.
9. https://rspo.org/.
10. https://josephpoore.com/Science%20360%206392%20987%20-%20Accepted%20Manuscript.pdf.
11. Rawlence, *Treeline*, p. 285.
12. https://www.nps.gov/articles/wildfire-causes-and-evaluation.htm.
13. https://www.statista.com/chart/23989/youth-mental-health-uk/.
14. s://www.belfasttelegraph.co.uk/news/kids-know-more-about-pokemon-than-wildlife/.
15. William Blake, *The Portable William Blake*, 1799, p. 30.
16. Simard, *Finding the Mother Tree*, p. 305.

CHAPTER 9: TREE HUGGERS

1. https://www.researchgate.net/publication/344720039_Radiocarbon_ Dating_of_the_Historic_Livingstone_Tree_at_Chiramba_Mozambique.
2. https://www.neh.gov/humanities/2012/novemberdecember/feature/ humboldt-in-the-new-world.
3. https://www.earlymoderntexts.com/assets/pdfs/bentham1780.pdf.
4. https://www.scientificamerican.com/blog/plugged-in/why-we-know-about-the-greenhouse-gas-effect/.
5. R W Emerson, *Nature*, James Munroe and Company, 1836, p. 15.
6. G P Marsh, *Man and Nature*, 1860.
7. https://www.britannica.com/topic/Sierra-Club.
8. https://www.nps.gov/yose/learn/historyculture/roosevelt-muir-and-the-grace-of-place.htm.
9. https://woodstocksanctuary.org/woodstock-blog/earthday2024.
10. https://www.newscientist.com/article/mg24933270-800-green-spaces-arent-just-for-nature-they-boost-our-mental-health-too/.

CHAPTER 10: SEEDS OF WISDOM

1. https://www.theguardian.com/books/2022/mar/21/the-big-idea-can-forests-teach-us-to-live-better-trees-model-suzanne-simard.
2. Reid & Lovejoy, *Ever Green*, p. 134.
3. https://wwf.panda.org/discover/knowledge_hub/where_we_work/ amazon/about_the_amazon/#:~:text=Not%20only%20does%20the%20 Amazon,river%20discharge%20into%20the%20oceans.
4. Carlos Peres & Maurício Schneider, 'Subsidized agricultural resettlements as drivers of tropical deforestation', *Biological Conservation* (2011), doi:10.1016/j.biocon.2011.11.011.
5. https://news.mongabay.com/2023/07/six-months-into-lulas-presidency-amazon-deforestation-is-dropping-rapidly/.
6. https://www.universityworldnews.com/post.php?story=2024040909 322761.
7. Reid & Lovejoy, *Ever Green*, p. 7.
8. Eduardo Kohn, *How Forests Think: Toward an Anthropology Beyond the Human*, University of California Press, 2013, p. 87.
9. https://theotherwise.net/files/issue1/TheOtherwise_Kohn&Ushigua.pdf.

10. https://branchoutnow.org/the-carbon-market-shell-game/.
11. David George Haskell, *The Songs of Trees: Stories from Nature's Great Connectors*, Viking Press, USA, 2017, p. 18.
12. Armstrong, *Sacred Nature*, pp. 6–7.
13. https://journals.openedition.org/emscat/3088.
14. https://www.smithsonianmag.com/smart-news/earliest-surviving-wood-sculpture-even-older-previously-thought-180977320/#:~:text=In%20 1997%2C%20Russian%20scientists%20carbon,to%20about%20 9%2C500%20years%20ago.
15. https://Indigenousfoundations.arts.ubc.ca/totem_poles/#:~:text=Most %20totem%20poles%20are%20made,grained%20and%20easy%20to %20carve.&text=Before%20a%20cedar%20tree%20is,in%20honour%20 of%20the%20tree.
16. https://hmh.org/library/research/genocide-of-Indigenous-peoples-guide/.
17. https://www.americanbar.org/groups/gpsolo/publications/gp_solo/2018/ may-june/standing-rock-case-study-civil-disobedience/#:~:text=The%20 group%20was%20leading%20the,the%20reservation%20and%20 Lake%20Oahe.
18. https://www.iwgia.org/en/sapmi.html.
19. Rawlence. *Treeline*, p. 55.
20. Rawlence, *Treeline*, p. 56.
21. https://www.usaid.gov/democratic-republic-congo/environment#:~:text= The%20planet's%20second%20largest%20tropical,being%20released%20 into%20the%20atmosphere.
22. https://minorityrights.org/communities/batwa-and-bambuti/.
23. https://minorityrights.org/they-came-to-purge-the-forest-by-force/.
24. Bruce Chatwin, *The Songlines*, Viking, 1987.
25. Davis, *The Wayfinders*, p. 150.
26. https://www.nature.com/articles/d41586-024-00693-6#:~:text=Fire%2Dstick%20farming%20involves%20 introducing,early%20in%20the%20dry%20season.
27. Davis, *Wayfinders*, p. 148.
28. Davis, *Wayfinders*, p. 152.

CHAPTER 11: HEALING FORESTS

1. https://www.jstor.org/stable/29791792?seq=4.

2. https://www.worldwildlife.org/stories/what-is-the-sixth-mass-extinction-and-what-can-we-do-about-it#:~:text=Unlike%20previous%20extinction%20events%20caused,been%20converted%20for%20food%20production.
3. https://www.cambridge.org/core/journals/review-of-international-studies/article/abs/global-tree-forests-and-the-possibility-of-a-multispecies-ir/DCB6DCD9B69DCAADD4D4AC0657FBC765#fn107.
4. https://www.ecologyandsociety.org/vol26/iss2/art6/.
5. Reid & Lovejoy, *Ever Green*, p. 9.
6. Dr Qing Li, *Forest Bathing: How Trees Can Help You Find Health and Happiness*, Penguin Life, 2018, p. 19.
7. Qing Li, *Forest Bathing*, p. 12.
8. https://digital.nhs.uk/data-and-information/publications/statistical/health-survey-england-additional-analyses/ethnicity-and-health-2011-2019-experimental-statistics/prescribed-medicines.
9. https://www.who.int/medicines/publications/traditional/trm_strategy14_23/en/.
10. https://www.ncbi.nlm.nih.gov/pmc/articles/PMC5662765/.

CHAPTER 12: LEAVES OF CHANGE

1. https://education.nationalgeographic.org/resource/isaac-newton-who-he-was-why-apples-are-falling/.
2. https://iucn.org/our-work/nature-based-solutions#:~:text=Nature%2Dbased%20Solutions%20address%20societal,nature%20at%20the%20same%20time.
3. https://forest.sabah.gov.my/usm/2020/publication/Annual%20Report%202020.pdf.
4. https://www.ox.ac.uk/news/2023-09-18-replanting-logged-forests-diverse-seedlings-accelerates-restoration-says-oxford#:~:text=Professor%20Hector%2C%20the%20lead%20scientist.
5. https://www.mdpi.com/1999-4907/13/10/1709.
6. https://arxiv.org/abs/2301.03354.
7. https://www.pnas.org/doi/10.1073/pnas.2004334117.
8. https://www.cam.ac.uk/stories/carbon-credits-hot-air.
9. https://www.worldbank.org/en/news/feature/2007/03/15/restoring-chinas-loess-plateau#:~:text=Replanting%20and%20bans%20on%20grazing,100%20million%20tons%20each%20year.

10. https://www.sciencedirect.com/science/article/pii/S0034425723002961.
11. https://airseedtech.com/what-we-do/.
12. https://dendra.io/blog/aerial-seeding-koalas-wwf-tff/.
13. https://earth.esa.int/eogateway/missions/biomass/description.
14. https://www.americanforests.org/article/trees-on-the-move/.
15. https://www.sciencedirect.com/science/article/abs/pii/
    S0921800915301725?via%3Dihub.
16. https://trilliontrees.org/.
17. https://ig.ft.com/one-trillion-trees.
18. https://forestdeclaration.org/resources/forest-declaration-assessment-
    2022/#:~:text=Overarching%20forest%20goals,-To%20be%20on&text=
    Globally%2C%20forests%20became%20more%20degraded,to%20
    halt%20deforestation%20by%202030.
19. https://www.london.gov.uk/publications/urban-greening-factor.
20. https://www.openaccessgovernment.org/land-surface-temperatures-
    heavily-influenced-by-urban-trees/124940/#:~:text=The%20team%20
    discovered%20that%20overall,found%20through%20the%20data
    %20collected.
21. http://web5.arch.cuhk.edu.hk/server1/staff1/edward/www/team/
    Publication/Tanya_in%20press.pdf.
22. https://is.muni.cz/el/1423/podzim2011/HEN597/um/Readings_Env_Psy/
    Kuo__F.E.__Sullivan__W.C.__2001_.pdf.
23. https://www.scientificamerican.com/article/nature-that-
    nurtures/#:~:text=All%20other%20things%20being%20equal,instead%20
    saw%20a%20brick%20wall.
24. https://newfaculty.uchicago.edu/page/marc-berman.
25. https://www.thelancet.com/journals/lancet/article/PIIS0140-6736(08)
    61689-X/abstract
26. Richard Walker, *The Green Grocer: One Man's Manifesto for Corporate
    Activism*, DK, 2021, p. 38.
27. https://ifcf.org.uk/.

CHAPTER 13: REFORESTING OUR MINDS

1. https://www.bbc.co.uk/news/uk-england-tyne-67077617.
2. https://rts.org.uk/article/growing-appeal-natural-history-tv.

3. https://smhp.psych.ucla.edu/pdfdocs/systemic/tipping%20point%20 summary.pdf.
4. https://scholarsarchive.byu.edu/wnan/vol68/iss4/8/.
5. Hanbury-Tenison, *Our Oaken Bones*, p. 50.
6. Isabella Tree, *Wilding: The Return of Nature to a British Farm*, Picador, 2018, p. 294.

Index